Groundwork 2
Junior Certificate Geography

The Educational Company

Preface

Groundwork provides a complete, well structured two volume Junior Certificate Geography course. Groundwork 1 covers the physical geography elements of the syllabus, while Groundwork 2 covers social and economic geography, map reading and aerial photographs.

- The language level is appropriate to Junior Certificate students. It is written in easy to understand language but without compromising the demands of the syllabus.

- Each double page spread deals with an aspect of a topic and is complete in itself. This enables the student to digest the topic in manageable 'bites'. Therefore, the student does not face such a daunting task when studying or revising the work. Each topic is framed with the key words at the beginning and key points at the end.

- The demands of the syllabus are rigidly adhered to throughout the books. The demands of the exam are also a priority in deciding on material covered in the books. The use of a large number and range of graphics to convey or support the information is in line with the growing emphasis on skills in the exam both at Junior Certificate and Leaving Certificate examinations. The contents of this book have also been organised so that students will be prepared for dealing with the new Leaving Certificate course.

- There is a great range of questions at the end of each chapter. The separation of questions from the text enables teachers to plan homework assignments effectively. These questions are based on the Junior Cert examination questions the content of the chapter and use a variety of techniques of examination.

This is a 'friendly text' with informal language, colourful illustrations and some humour, which will appeal to young students.

Contents

1 Population Studies – Numbers of People in the World

Key words

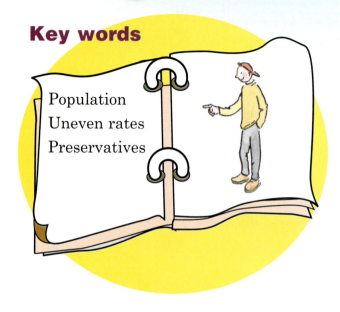

Population
Uneven rates
Preservatives

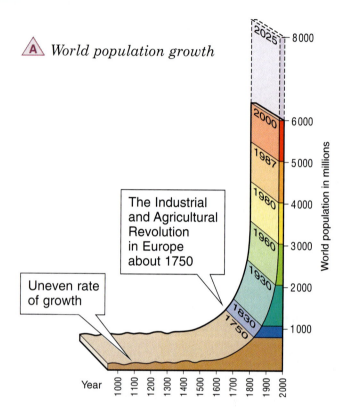

World population growth

2025 — 8 000

2000 — 6 000

1987 — 5 000

1980 — 4 000

The Industrial and Agricultural Revolution in Europe about 1750

1960 — 3 000

1930 — 2 000

Uneven rate of growth

1830 1750 — 1 000

World population in millions

Year 1000 1100 1200 1300 1400 1500 1600 1700 1800 1900 2000

World Population

Today there are over 6 billion people living in the world, but for a long time the total world **population** was less than 1 billion. A billion is a thousand million, a figure almost too big to imagine.

1,000,000,000 = 1 billion

To give you some idea of how big this figure is, it would take you eleven days to count to 1 million if you allowed one second for every number you counted.

The story of population growth

Let's look at the story of how population numbers have grown since humans first appeared on the planet.

World population grew at an **uneven rate** up until the 1700s. There were times when the population increased and other times when it fell rapidly.

The drop in population numbers happened because of events such as wars, famines and plagues. People were unable to survive illnesses and diseases. There were few medicines. The Black Death (spread by rats!) in the 1400s wiped out huge numbers of people in Europe. People could do little about it.

There were times in the past when the harvest failed and people went hungry and died. There were no **preservatives** in food and no freezers to carry people over in times of food shortages. There were no newspapers or television to tell you that a famine was happening in another part of the world. So, there was no organised relief for famine victims.

Rapid population growth

Up until 1750, the total population of the world remained under 1 billion.

After 1750, the number of people in the world began to increase very rapidly. The increase happened mainly because people began to live longer. This had to do with improvements in farming and in medicines. At this point the number of people being born was much greater than the number of people dying. The population began to grow very fast.

This increase in population numbers has continued right up to the present. The increase has been particularly rapid during the twentieth century. This increase in the numbers of people on the earth is mainly happening in poor countries.

Between 1750 and 1930 the population of the world increased from 1 billion to 2 billion. However, between 1930 and the present, a further 4 billion was added to the figure which has reached 6 billion.

Rapid population growth in the developing world

High birth rates swell population numbers in poorer countries of the world

Key points

- The **population** of the world increased very slowly up until 1750.
- Improvements in medicine, in farming and in the use of food **preservatives** caused a fall in the number of people dying.
- There has been a rapid increase in world population since the beginning of the twentieth century.

Population Cycle Model – Changes in Time

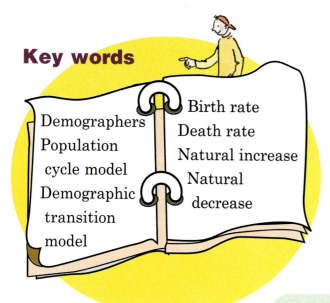

Key words

Demographers
Population
cycle model
Demographic
transition
model

Birth rate
Death rate
Natural increase
Natural
decrease

A model gives a simple picture of how things can turn out. In this case the **population cycle model** gives an idea of what happens to population numbers as a country moves through different stages to become more developed.

Before looking at the diagram you need to learn some key words.

The world population numbers have increased rapidly over the past century. The greatest increase is happening in the poorer countries of the world.

Demographers (the experts on population) have noticed the following pattern:

- As a country starts to develop, its population increases very rapidly.
- As it develops further its population growth begins to level out or may even fall in size.

This pattern is called the **population cycle model** and it can be shown on a diagram. Another name is the **demographic transition model**. It is based on what happened in Europe when it started to develop in the nineteenth century.

- Birth rate

  The **birth rate** is the numbers of babies born per year for every 1,000 people in the total population.

- Death rate

  The **death rate** is the numbers of people dying per year for every 1,000 people in the total population.

- Natural increase – natural decrease

  Natural increase means the difference between the birth rate and the death rate. When the birth rate is higher than the death rate there will be a natural increase (the population will increase in size). When the death rate is higher than the birth rate, there will be a **natural decrease** (the population will fall in size).

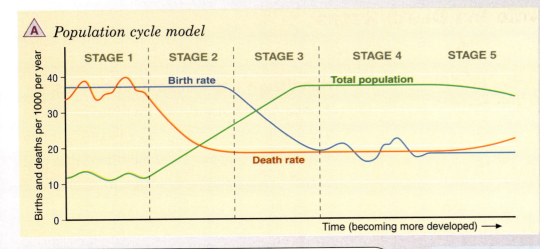

A *Population cycle model*

STAGE 1 | STAGE 2 | STAGE 3 | STAGE 4 | STAGE 5

Births and deaths per 1000 per year

Birth rate
Total population
Death rate

Time (becoming more developed) →

The population cycle model

- Stage 1 is the stage when a country is undeveloped i.e. poor. There are high **birth rates** and high **death rates**. The number born is matched by the number dying, so the population stays the same or increases only very slowly. There are probably no countries in the world at this stage now.

- Stage 2 is the stage when the country begins to develop. **Birth rates** are high but **death rates** begin to fall. There is a big gap between birth rates and **death rates**, so the population will increase rapidly. Mali in Africa is at this stage.

- Stage 3 is the stage when the economy is developing well. **Birth rates** begin to fall and **death rates** are low. The gap between the **birth rate** and **death rate** is getting smaller, so the population does not grow as rapidly. Brazil in South America is at this stage.

- Stage 4 is the stage when a country has become developed. Both **birth rates** and **death rates** are very low. In this case, the gap between the two rates is very small, so the population does not increase very much. Ireland is in this stage of its development.

- Stage 5 is the stage when the economy is even more developed. The **birth rate** is actually lower than the **death rate** so the population decreases. Countries like Germany are now in this stage.

Brazil, a developing country

- The population of Brazil is approximately 159.5 million.
- The birth rate is 20/1,000 people per year.
- The death rate is 9/1,000 people per year.
- Therefore, the population has a natural increase of 11/1,000 people per year.

Germany, a rich country

- The population of Germany is approximately 82 million.
- The birth rate is 9/1,000 people per year.
- The death rate is 11/1,000 people per year.
- Therefore, the population has a natural decrease of 2/1,000 people per year.

Key points

- Population changes as a country goes through different stages of development.
- This is shown on the **population cycle model**.

Changes in Birth and Death Rates

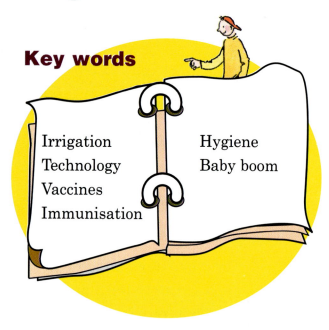

Key words

Irrigation
Technology
Vaccines
Immunisation

Hygiene
Baby boom

Birth rates and death rates change as a country becomes more developed.

Food supply

As a country becomes more developed, its food supply improves. Machines make it easier to sow and harvest crops.

Irrigation (watering the land) improves the supply of food. Food with added preservatives is widely available and lasts longer. That means there will always be a supply of food. This leads to a lower death rate.

Improved technology

There is a lot of **technology** in our lives today. A water treatment plant gives us clean water. Technology in our hospitals has brought us heart machines and X-ray machines. Technology has made it possible to research diseases and come up with cures. All of these machines reduce the death rates.

Healthcare and education services

A good healthcare system will bring down the death rate. When you are examining a country to see how developed it is, it is useful to look at the average number of people for every doctor.

Germany	367 persons per doctor
Ireland	632 persons per doctor
Brazil	844 persons per doctor

These figures give us an idea of the healthcare available. It is usually the case that the more doctors, nurses, hospitals and medicines there are in a country, the healthier the population. **Vaccines** and antibiotics control illnesses. Good healthcare lowers the death rate.

Education teaches people how to look after themselves. Schools can introduce good health programmes. Advertisements can teach people about health issues such as **immunisation** and **hygiene**. They may also inform people about family planning. This allows people to have smaller families causing a lower birth rate.

In this way, health education can lead to lower birth and death rates.

Healthcare influences birth and death rates

Role of women in society

In some societies the main role of a woman is that of a mother. Women marry young and have many babies. They rarely choose a career outside the family. In these countries men expect to father a large number of children. This leads to a high birth rate.

As a country develops women are more likely to find work outside the home. This leads to a lower birth rate.

Babies? Not Part of the Plan

As Spain's fertility dwindles to 1.13%, two Spaniards tell why they have decided to stay childless.

Lola Rodriguez, 33, has been with her boyfriend, Jon, for 11 years and has never felt the need to have children. 'I just don't like the thought of being in the role of a mother,' she says. 'It doesn't fit with my plans or lifestyle.' She has a good career and a good social life, and is happy with how things are.

Cristina Mendia, 50, has other reasons for remaining childless. 'The main issue is whether we have the right to have our own children when the world is full of children that need care: should we add to the population when so much should be done to improve the lives of these children who are already here?'

War

War has an effect on the birth rates and death rates in a country. Death rates are high when there is a war. The birth rate will fall when young men are away in times of war. A **baby boom** may happen on their return. This happened in Germany during and after the First and Second World Wars.

Key points

- Birth rates and death rates change as a country becomes more developed.
- Food supplies, **technology**, education and healthcare services have an effect on birth and death rates.
- The role of women in society influences the birth rates.
- Wars affect both the birth and death rates in a country.

Population Increase – The Rate of Change

Key words

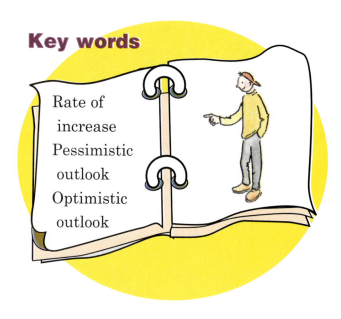

Rate of increase
Pessimistic outlook
Optimistic outlook

The rate of population change has to do with how fast populations are growing. Look at the table below. It shows how the population is changing in some countries. The figures given for 2050 are predictions made by demographers, experts on population numbers.

Countries	1950	2003	2050	Type
China	562	1,289	1,394	Developing
India	370	1,069	1,628	Developing
Japan	84	128	101	Developed
Germany	68	83	73.6	Developed
Brazil	53	176	221	Developing
France	42	60	61	Developed

A *Population in millions*

Brazil

Look at the figures for Brazil. In 1950 its population was 53 million. By 2003 its population had increased to 176 million. This was an increase of over 230%. Its population was increasing at a very fast rate. By 2050 experts believe that the population will have increased to 221 million. While this shows an increase of 45 million, it is less than the increase of

83 million between 1950 and 2003. This tells you that the **rate of increase** is slowing down in Brazil.

Brazil is a developing country and its figures are typical of such poor countries. Overall, they show a high rate of population increase.

Germany

Look at the figures for Germany. In 1950 its population was 68 million, larger than that of Brazil. By 2003 its population had grown to 83 million. Notice that Brazil, with a much faster growth rate of population, now had double the numbers in Germany. By 2050, over a period of about 50 years, the population of Germany is expected to fall.

Germany is a developed country and its figures are typical of such rich countries. They show a low rate of population increase, or even a decrease.

The rate of increase

Knowing the rate of increase is important because you can predict what figures are likely to be in the future.

Some people believe that the population of the world will continue to increase very rapidly – a pessimistic outlook. Others believe that the increases will not be so great – an optimistic outlook. Let's look at each of these outlooks.

Pessimistic outlook

Some people hold the view that by the year 2050 there will be a massive 25 billion people in the world. Most of these people will live in developing countries.

This is a worrying outlook because food production could not keep pace with the growth in population numbers. It would mean there would be huge food shortages and that famine would be likely to occur. This of course is a very **pessimistic outlook**. Pessimistic means that the future would look very bleak indeed.

Optimistic outlook

There are others who hold a more **optimistic outlook**. They look at the brighter side of things. Their view is that the numbers will not increase rapidly. Indeed, they believe that population numbers will remain steady. Instead of thinking that the number will reach 25 billion, they think it will peak at 8 billion in 2050. It is 6 billion at present.

The optimists predict that the birth rate will drop because education and healthcare are now more widely available to people. In other words fewer babies will be born. People all over the world will have smaller families.

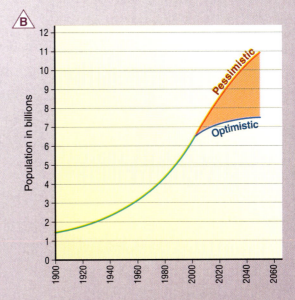

Predictions of population growth

Key points

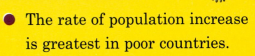
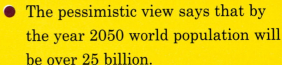

- The rate of population increase is greatest in poor countries.
- The pessimistic view says that by the year 2050 world population will be over 25 billion.
- The optimistic view says that future population will peak at 8 billion in 2050.

REVISION EXERCISES

Write the answers in your copybook.

1 Look at the chart.

(a) When did the population reach 3 billion?

(b) Discuss **two** problems that may result from the big increase in world population.

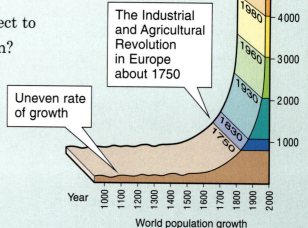

The Industrial and Agricultural Revolution in Europe about 1750

Uneven rate of growth

World population growth

2 Which of the following would you **not** expect to see in a rapidly growing human population?

● As many 20 year olds as 2 year olds

● As many females as males

● More 15 year olds than 25 year olds

● Large families

● More 4 year olds than 14 year olds

3 Birth rate per 1,000 per year = 13

Death rate per 1,000 per year = 12

The figures above for birth rate and death rate in a country lead to:

● The population increasing greatly

● The population decreasing rapidly

● The population increasing slowly

● The population decreasing slowly

4 The population cycle refers to:

● The way a population increases

● The way a population levels out

● The way a population changes as it becomes more developed

5 Calculate the rate of change in the population in China, India and Japan.

Population in millions

Countries	1950	2003	2050	Rate	Type
China	562	1,289	1,394		Developing
India	370	1,069	1,628		Developing
Japan	84	128	101		Developed
Germany	68	83	73.6	22%	Developed
Brazil	53	176	221	232%	Developing
France	42	60	61		Developed

6   During 1999 the world's population was about 6 billion. The table below shows the distribution of population between developed and developing countries from 1930 until 1999.

World Population 1930–99 (Billions)			
Year	Developed	Developing	Total
1930	0.3	0.7	1.0
1950	0.8	1.7	2.5
1975	1.0	3.0	4.0
1999	1.1	4.9	6.0

With reference to the figures in the table above, describe and explain **two** changes in the world's population.

7   Match each letter in column X with the number of its pair in column Y.

X	Y	ANSWER
A High birth rate, high death rate	1 Stage 3	A =
B Lower birth rate, lower death rate	2 Stage 1	B =
C Decreasing birth rate, low death rate	3 Stage 2	C =
D High birth rate, lowering death rate	4 Stage 4	D =

8   Give **two** reasons why the population of Europe changed after 1750 and explain your answer.

9   Describe, using examples, two reasons why the birth rates in a country may change.

10 Wars and health education often change the death rate in a country. In the case of one country that you have studied show how this is true.

11 The populations in Brazil and Germany are increasing at a different rate. Explain how this rate is calculated.

12 People have very different outlooks as to how high world population is likely to grow. One group's view is described as the pessimistic outlook, the other group's as the optimistic outlook. Explain clearly the difference between these two outlooks.

2 Population Distribution and Density – World Level

Key words

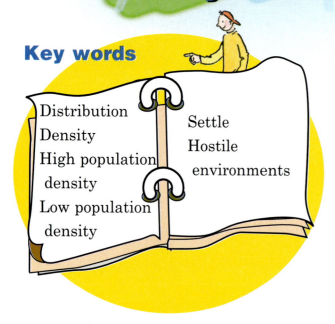

Distribution
Density
High population
  density
Low population
  density

Settle
Hostile
  environments

The term population **distribution** refers to the way people are spread out across the earth's surface.

Look at the map showing the distribution of population in the world. You will see that there are huge areas that have few people living in them. There are other areas that have many people living in them. We call this an *uneven* pattern of population distribution.

Population **density** has to do with how crowded a place is. Areas that have many people living in them are said to have a **high population density**. Where there are few people, we call this **low population density**.

To understand this, imagine a city like New York. There is high population density here. On the other hand, picture an area of farmland where settlements (houses) are spaced out far from each other. There is low population density here.

We can calculate population density. It can be shown as the average number of people living in a square kilometre. It is worked out by dividing the population total by the area of the country.

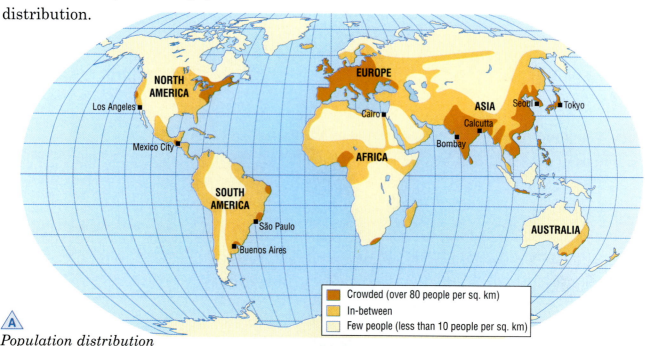

Legend:
- Crowded (over 80 people per sq. km)
- In-between
- Few people (less than 10 people per sq. km)

A
Population distribution

- Ireland has a population of 3,950,000
- The area of Ireland is 70,283 km²
- 3,950,000 divided by 70,283 is 56
- The density of population in Ireland is 56 people per sq. km

People **settle** in some places and avoid others for good reasons.

- People avoid living in **hostile** (harsh) **environments.**
- People will settle where they can make a good living.

Areas that are hostile places to live

B *Harsh desert landscape*

A hostile environment is a place where the natural environment makes it hard for people to survive there. For example:

- People avoid living in deserts where there is a shortage of water. The Sahara has very few settlements.
- People avoid living in very cold environments because it is hard to grow food there. Northern Canada, Norway and Russia have low population density.

Harsh, cold landscape of the North

Areas that are good places to live

- There are high population densities on lowlands and in river valleys where there is a water supply and fertile soils. Many people live in the valleys of the River Rhine, Germany and the River Ganges, India.

High population density in the Rhine river valley

- People live near coasts where it is possible to trade. The coastline of south-east Brazil and eastern Ireland show these high densities.

High population in the coastal city of Dublin

Key points

- Population **distribution** is about the spread of people in an area.
- Population **density** is about how crowded a place is. It is presented as the average number of people in a square kilometre.
- People in general avoid **hostile environments** such as deserts or cold places.
- People **settle** on lowlands, in river valleys or near coasts.

Case Study: Population Density in Italy

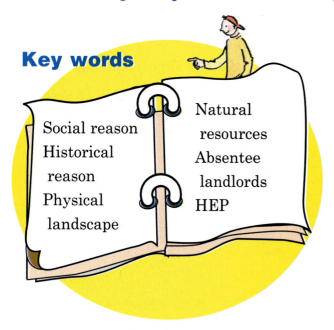

Key words

Social reason
Historical reason
Physical landscape

Natural resources
Absentee landlords
HEP

There are good reasons for the differences in the spread of population. Some are **social** and **historical reasons**, others are to do with the **physical landscape** and the **natural resources** that occur in the region.

Differences in population density within countries

When you look at a map showing population distribution and density, you are looking at a very general picture of where people live in that country. When you look more closely, you will see that there are differences *within* a country that don't show up on a large map.

Let's look at population differences within the country of Italy. Look at map .

Italy: differences in population distribution and density

Some areas of Italy have many people and some have very few. The most crowded places are in Northern Italy. The least crowded places are in Southern Italy.

- Northern Italy's population density is 230 per km².
- Southern Italy's population density is 160 per km².

Northern Italy

Social and historical reasons

Many people have settled in Northern Italy for the following reasons:

- Northern Italy is crossed by one of the oldest trade routes, the silk route from China. Many towns grew as a result of this passing trade.
- Northern Italy is well linked by road and rail with the rest of Europe. With good links, trade and towns grew.

Physical landscape and natural resources

- In Northern Italy the **physical landscape** is flat, with fertile alluvial soils which are good for farming.
- Warm temperatures and summer rain suit the growing of wheat, fruit and vegetables.
- **Natural resources** such as **HEP** (power from fast-flowing water) and natural gas have given rise to many industries in Northern Italy.

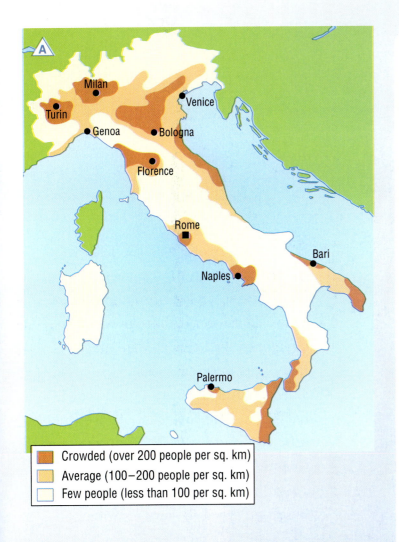

Crowded (over 200 people per sq. km)
Average (100–200 people per sq. km)
Few people (less than 100 per sq. km)

Southern Italy

Social and political reasons

- In the past, **absentee landlords** owned most of the land. They let the land to poor tenant farmers, who could not afford to irrigate (water) the land or use fertilisers. The land could not offer a good living to the people. They moved away.

- In the past, many of the people from Southern Italy did not have basic reading and writing skills. Businesses preferred to set up in Northern Italy. This caused people to leave the area to find work.

Physical landscape and natural resources

- In Southern Italy, the **physical landscape** is mountainous. With steep slopes and thin, infertile soil it is difficult to farm. It is also hard to build roads.

- Southern Italy does not have any **natural resources** such as **HEP** or natural gas. It is therefore hard to attract factories into this area. When there are few jobs people will go elsewhere to work and live.

Changes

Changes are being made in Southern Italy. The government has broken up the large farms and given farmers pieces of land that will offer them a good living. They have given grants to businesses to set up here. They have trained workers in new skills. All of this should help to stop the flow of people from the South. This should lead to higher population densities in the future.

Key points

- Differences in population distribution and densities occur within countries.
- Northern Italy has a high population density.
- Southern Italy has a lower population density.
- People will live in places where the landscape and the resources offer a good living.

Case Study: Brazil

Key words

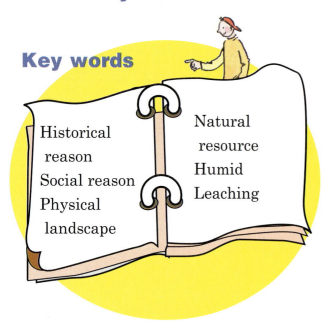

Historical reason
Social reason
Physical landscape

Natural resource
Humid
Leaching

Brazil is a country in South America. It is about 103 times the size of Ireland. It is the fifth largest country in the world. Five hundred years ago, Brazil was inhabited only by tribes of Indians. Then the Portuguese came and ruled the country as a colony for over 300 years.

Brazil has a population of almost 165 million – 50% of the people who live there are under the age of 20. The population is growing very fast.

In Brazil, people are not spread out evenly throughout the country. There are areas with very few people and other areas that are really crowded.

Let's look in more detail at one area of high population density and one area of low population density.

Population distribution in Brazil

South-east Brazil is very crowded, it is densely populated. The Amazon Basin in north and west-central Brazil has few people, it is very sparsely populated.

South-east Brazil
Social and historical reasons

- South-east Brazil was the area first colonised by the Portuguese. They came by sea and settled along the coast. These coastal areas were developed and became the most heavily populated areas.
- The Portuguese built a good road network along the coast. These links attracted more people and industries. Coastal cities like São Paulo (pop: 18 million) and Rio de Janeiro (pop: 6 million) became very crowded.

Physical landscape and natural resources

- People settled on the narrow, flat stretches of land along the coast. The fertile soils here are ideal for farming.
- Warm temperatures and plentiful rain give excellent conditions for crops like coffee.
- **Natural resources** such as gold and iron ore are found in the south of Brazil. These attract industries. There is also a good natural harbour at Rio de Janeiro which brings trade and settlers.

Crowded (over 25 people per sq. km)
In-between (7–25 people per sq. km)
In-between (1–6 people per sq. km)
Few people (less than 1 per sq. km)

Amazon River Basin

Social and historical reasons

- The Amazon River Basin is very hot and **humid**. The Portuguese saw this area as unhealthy and very remote. Very few people were attracted to living here.
- The government was slow to invest in building roads through the dense forests. Few roads link this area to the coast. As a result, few industries and towns set up here.

Physical landscape and natural resources

- Dense forests make travel difficult through the Amazon River Basin. For this reason, it is difficult to make use of the **natural resources** of the area. It is hard to make a living here.
- High rainfall amounts cause **leaching** of minerals in the soil. The resulting poor soils make farming difficult. Fewer than one person per square kilometre can be supported in this hostile environment.

- In recent years the government has tried to encourage others to settle here. They built the Trans-Amazonian highway to open up the interior of the country.

Trans-Amazonian highway

- They offered settlers free land on which to graze cattle.
- Mining companies arrived to dig out the minerals found here. There has been a small increase in population densities in recent years.

Key points

- Brazil's population is unevenly distributed.
- Most people live in the south and south-east of the country.
- This is, in part, due to the many trading links that developed.
- Low densities of population are found in the Amazon rainforest.
- Recent efforts have been made by the government to tempt people to move into the Amazon Basin.

Ireland: Differences in Population Densities

Key words

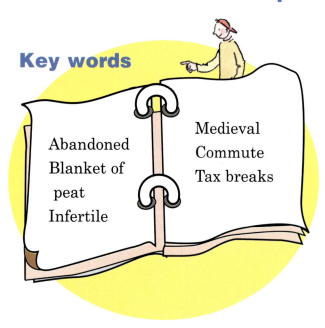

Abandoned
Blanket of
peat
Infertile

Medieval
Commute
Tax breaks

The West of Ireland is an area with few people. It has a history of people moving away. Let's look in more detail at the reasons for low population density in the West.

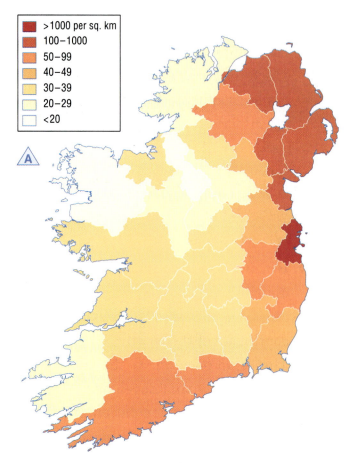

▮	>1000 per sq. km
▮	100–1000
▮	50–99
▮	40–49
▮	30–39
▮	20–29
▯	<20

Ⓐ

Differences in the population density within Ireland

The West of Ireland
Social and historical reasons

● In the past, when a farmer died, the land was divided among his sons. In 1841 almost half the farms in the West were smaller than five acres. Small farms could not offer a good living to people, so many **abandoned** their farms and left the area.

● People from the West moved to the Dublin region in search of work. Here, there were jobs in government offices, hospitals and factories. This loss of people from the region led to low population densities.

An abandoned farmhouse in the West of Ireland

Physical landscape and natural resources

● It is difficult to make a living from farming in the West. A **blanket of peat** covers mountains. Rainfall is high. The soils are thin and **infertile**.

● Up until recent times very little money was spent improving the roads. Few industries were set up for this reason. People had to move to find work.

The Dublin region

The population of the Dublin area has almost doubled over the past 50 years. Nearly 30% of the country's population now live there. The population is unevenly spread throughout this area.

Social and historical reasons

- Dublin is a **medieval** settlement. The city began in the area around Christchurch. In early times buildings were crowded close to the heart of the city. The map shows that it is most crowded near the centre of Dublin.

- Over time more people moved into the city. Transport improved and the city spread out. Wealthy people moved to the coast and **commuted** to the city by train or car. There are still high population densities along the coast and the main roads leading into the city.

- Government planning has led to high population densities in certain parts of the city. Lucan, Clondalkin, Tallaght and Blanchardstown form a cluster of new towns on the edge of Dublin. Many new houses and apartments have been built close to the city centre with the help of generous tax breaks.

Physical landscape and natural resources

- The fertile land to the north of the city is used for farming rather than housing. Dublin Airport needs lots of land. It is also a very noisy place.

- Dublin City has a long history of industry. People settle where there is work, giving rise to higher population densities towards the city centre.

- The mountains lie to the south of Dublin. It is costly to build on steep slopes.

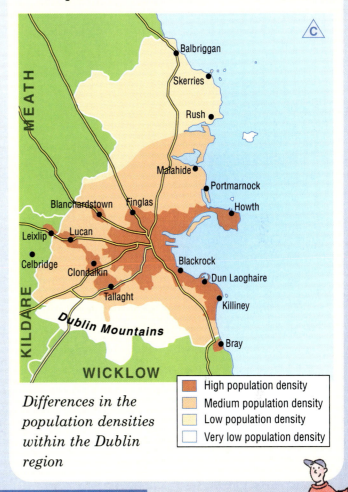

Differences in the population densities within the Dublin region

Legend	
■	High population density
■	Medium population density
■	Low population density
□	Very low population density

Key points

- The West coast of Ireland has a low population density.
- People moved away from the West because they could not make a good living there.
- The centre of Dublin and areas along the coast have a high population density.
- The south and north of the city have low population densities.
- Other areas, have high population densities because of government planning.

REVISION EXERCISES

Write the answers in your copybook.

1 A place is said to have too many people when:
- The population is high
- There are many big cities
- The population is growing
- There are not enough resources for people to survive

2 In which continent is the country of Brazil located?
- Asia
- Europe
- South America
- North America

3 The density of population means:
- Where people live
- The average number of people in every square kilometre
- The average number of people in a county
- The total population in a country

4 The population density of Ireland is:
- 56 people per square kilometre
- 4 million people
- 3.7 million people
- 454 per square kilometre

5 Match each letter in column X with the number of its pair in column Y.

X	Y	ANSWER
A High population density	1 Mountainous areas	A =
B Population density	2 The spread of population	B =
C Low population density	3 How crowded a place is	C =
D Population distribution	4 Fertile river valleys	D =

6 Look at the world map on page 12 showing the distribution of population. Which of the following statements are true?
(a) People are not spread evenly over the world.
(b) People are spread evenly over Asia.
(c) Some places in the world are more crowded than others.
(d) Europe is a crowded continent.
(e) South America is the most crowded continent.
(f) Africa is a continent with few people.

- a, c, e, f
- a, c, d, f
- a, b, c, d
- b, c, e, f

7 Some areas of the world are more crowded than others. Using named examples, explain **two** reasons why this is so.

8 Most Italians live in cities. One of those cities is Milan. Write a paragraph which explains the reasons why so many people like to settle in or near this city.

9 The population of Brazil is unevenly distributed. Show why this pattern has developed using the headings:
- Type of land
- Natural resources

10 The West of Ireland has a low population density. Explain the historical reason for this pattern.

11 Name two areas of Dublin which have a high population density. In the case of **one** of these areas explain why so many people have chosen to settle there.

12 Give **one** detailed reason why there is a low density of population in North Co. Dublin.

3 Population Pyramids

Key words

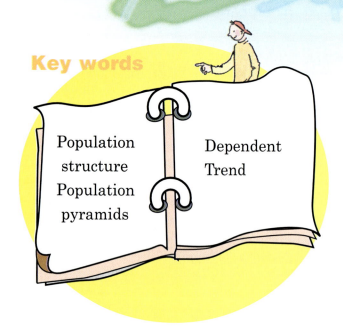

Population structure
Population pyramids

Dependent
Trend

Population Structure

The sex structure of populations looks at how the number of males in the population compares with the number of females.

Population structure is also about the age make-up of the population. This shows how the percentage of old people, for example, compares with the percentage of children in the population. This is its age structure.

Think about the age/sex structure of your school or your community. How does the percentage of males compare to that of females or older people to younger?

Population pyramid A

A graph called an age/sex **pyramid** gives an image of how the population is made up *at a particular time*. It shows the percentage of the population in each five-year age group. It also shows you how the percentage of males compares with the percentage of females in each age group. This is how the graph is built.

The pyramid is made up of bars running from left to right (horizontal bars). Each bar stands for a block of five years. The length of the bar tells you the percentage (%) or the number in each age group.

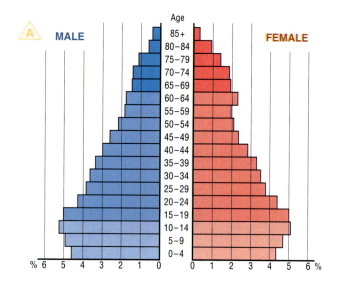

Look at this on the graph. Males are shown to the left of the graph. Females are shown to the right.

The population pyramid shows you:
- The age/sex structure at a particular time.
- An idea about the age/sex structure in the future.
- A picture about what happened to the population in the past.

This information is of great interest to governments as it helps them in planning for the future.

Let's take each one of these and look at it in a little more detail.

The present make-up of the population

One of the things the pyramid shows you is the percentage of the population in the age brackets 16–65. This is the working age group. They pay taxes to the state, i.e. money that is used to pay for services for themselves and people in the **dependent** age brackets (those between 0–16 and over 65). These groups are called dependent because they rely on the support of the working group.

The government needs to know how much tax it is likely to collect and what it needs to spend it on. Knowing the age structure of the population makes this possible.

Future trends

The **population pyramid** gives important information about future **trends**. Where is this population going? What changes are happening?

To understand the idea of **trend** consider the following example.
- How does the percentage of children between 10–14 compare with the percentage aged between 0–4? Does it look like families are getting smaller?

The government will pay close attention to the figures and try to work out what the population make-up will be like in future years. There is no point extending a secondary school, for example, if numbers in the primary school are falling.

Governments will examine the age structure to see if the population is getting older.
- Will there be far more old people than young people in the population in the future?
- Do people need to work beyond the age of 65 years?
- Are there enough workers under 65 to pay pensions for these older people?

Noticing these trends is very important. The government wants to prepare for the services that will be needed in the future.

Key points
- **Population pyramids** show the age/sex make up of a population.
- Governments take note of **trends** in the **population pyramid** and plan for the future.

A Closer Look at Population Pyramids

Key words

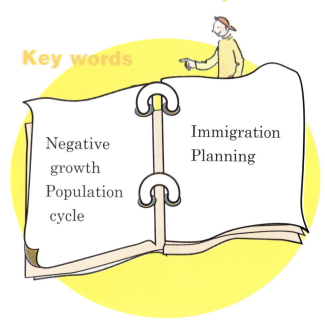

Negative growth
Population cycle

Immigration
Planning

There are three basic types of population pyramids.

- One pyramid shows a population which is growing very rapidly e.g. Brazil.
- A second shows a population which is actually falling. This is **negative growth** e.g. Germany.
- A third shows slow population growth e.g. Ireland.

Brazil – rapid population growth

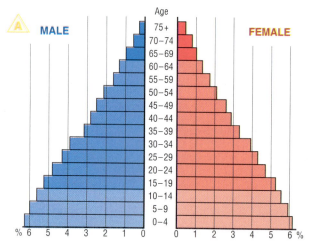

Notice these details in the population pyramid for Brazil:

- It has the shape of a triangle.
- The base of the pyramid is wide i.e. there are a large number of children under 15 years. This means it has a high birth rate.

- The pyramid tapers or narrows very steeply to the top. Few people survive into old age.
- This is a developing country as there is a high birth rate and a fairly high death rate.

Brazil is at stage 3 of the **population cycle**.

Germany – negative population growth

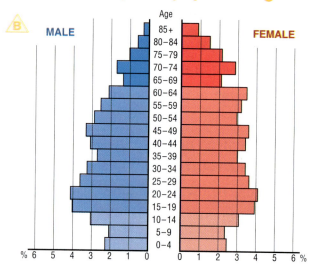

Notice these details in the population pyramid for Germany:

- This pyramid has a narrow base i.e. there is a small percentage in the 0–4 age bracket. This means there is a low birth rate.
- This pyramid bulges in the middle. There is a high percentage in the 15 plus age group. This may be due to **immigration**.
- The pyramid does not narrow steeply. It is wide as far as 70 years. This means that Germans survive into old age.
- There are more women than men in the older age groups.

Germany is at stage 5 of the population cycle.

Ireland – slow population growth

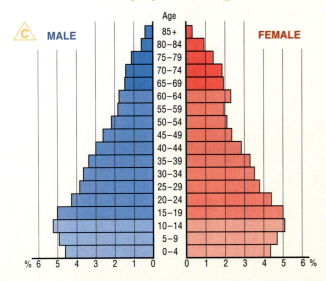

Notice these details in the Irish population pyramid:

- The population pyramid here is quite square in shape. This shows there is a low death rate. Many people survive into old age.
- The base of the pyramid (0–4) is slightly narrower than the next bracket up (5–9). There are signs of a decrease in the birth rate.

Ireland is at stage 4 of the **population cycle.**

Local study

Let's take a local area for study. There may be differences in population structure between one local area and another. Think about your own local area. Is there a new housing estate with a young population and an older area where there are a lot of older people?

Area 1

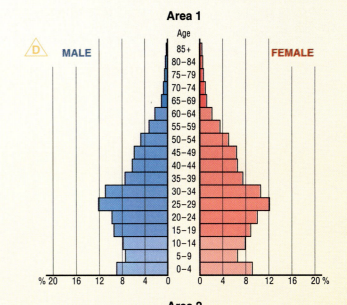

Area 2

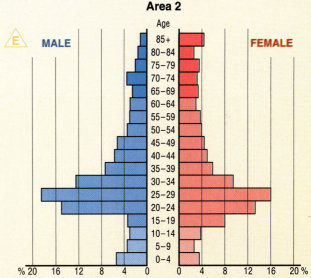

Notice the following details in pyramids D and E :

- Area 1 has a higher percentage of young children.
- Area 2 has a high percentage of young adults, 20–30 years old.
- Area 1 has a lower percentage of people over 65 years old, than area 2.

Key points

- Population pyramids show us three types of growth in a population: rapid, slow and negative.
- Within a local area there may be differences in the population structure.
- Information on population structure is needed for future **planning** by governments.

REVISION EXERCISES

Write the answers in your copybook.

1   Study the diagram. What is the
    percentage of those under 15?
    ● 23.9%
    ● 16.9%
    ● 8.6%
    ● 8.3%

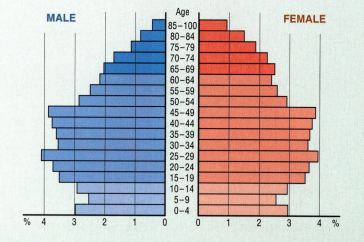

2   Study the pyramid given in question 4.
    In your copybook write the correct answer for **each** of the
    statements below.
    (a) The narrow base shows a low **birth/death** rate.
    (b) There are more **males/females** over the age 70.
    (c) The graph shows a **developing/developed** country.

3   Give the meaning of the following phrases:
    ● Birth rate
    ● Death rate
    ● Natural increase

4   The diagram opposite shows the
    population of a European city.
    Study the pyramid and
    answer the following questions:
    (a) What is the percentage of
        males in the 0–14 age group?
    (b) Complete the pyramid by
        drawing on the diagram
        opposite the following bar –
        Males 60–74 (6%).

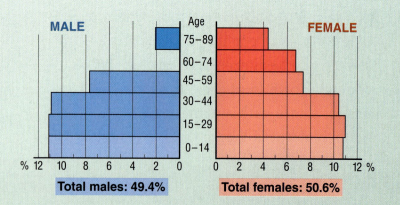

5 Study this age/sex pyramid which shows the population of part of a large city. The population shown by this age/sex pyramid is likely to be found in:
- A suburb with new houses
- The inner city
- The area around a university
- The central business district

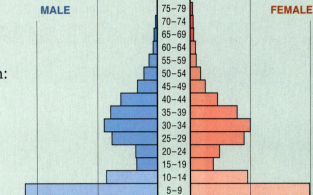

6 A population pyramid can be divided into 'the dependent group' and 'the productive (or working) group'. Explain each of these terms. In which group would you put a girl of 6 years old and a man of 71 years old?

7 Explain what use an age/sex pyramid such as the local one given above would be to:
- The principal of a new school
- A local bus company
- The owner of an old folks' home in the area

8 Examine the pyramids representing the populations of Brazil and Germany.
Describe two differences between them. In your answer refer to birth and death rates.

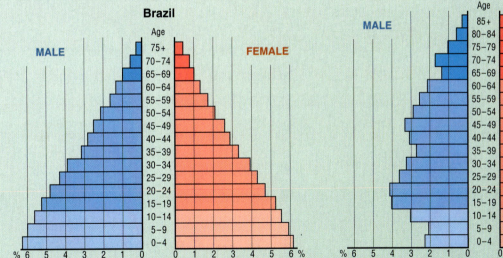

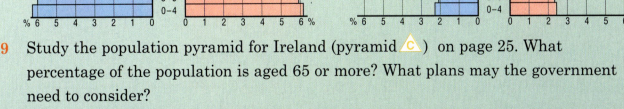

9 Study the population pyramid for Ireland (pyramid c) on page 25. What percentage of the population is aged 65 or more? What plans may the government need to consider?

10 Write down **one** effect that **each** of the following factors may have on Ireland's population pyramid:
- War
- The opening of a large factory
- Return of many retired Irish people who had worked in Britain

4 Population Density: Its Effects

Key words

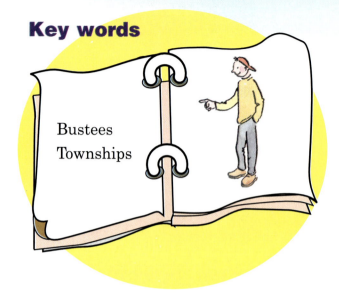

Bustees
Townships

Calcutta: High Population Density

Calcutta, in India, is a poor area with high population density. It is one of the most crowded cities on Earth. It is built along the banks of the Hoogley River. Calcutta is an important port, serving north-east India, Nepal and Bhutan.

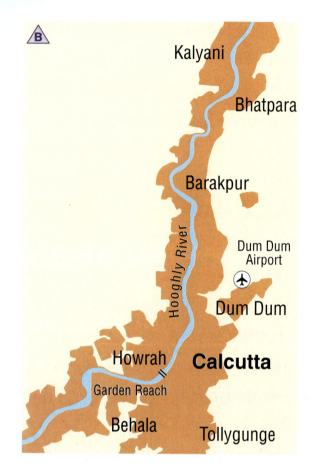

Calcutta has a population of over 10 million and this figure is growing by 28% every year. The rapid increase in its population is causing many problems for the city and its people. The problems are:
● Overcrowding
● Lack of open space
● Pollution
● Shortage of clean water.

Overcrowding

There is a serious shortage of housing in Calcutta. Many families have no homes at all and sleep on the pavements. Over a quarter of a million people are said to sleep in the open, covered only by bamboo, sacking or newspapers.

A further 4 million people live in slums called **bustees**. Most of the people living in the slums are poor migrants, who have come from other places in India, with a dream of a better life.

Their homes are built from cardboard or wattle, with tiled roofs and mud floors. These houses don't protect the people from the heavy monsoon rains.

Dwellings are packed closely together and are separated by narrow alleys. Inside each home there is usually only one small room in which the family, often up to eight people, live, eat and sleep. Many houses have no running water or toilets.

The government has built new housing areas called **townships** at the edge of the city. There is such a rapid growth in population and such a shortage of land for housing that there are simply not enough houses. Many people still live in slum conditions.

Several new planned townships have been built on the eastern fringes of the city.

Shortage of open space

Calcutta is a very built-up area. The demand for land is so great that very little land is left as open space. The biggest park 'The Maidan' is under threat as new space for housing is urgently needed.

Often, the only space for children to play is car-parking space. Even in the richer areas this is the case.

Pollution

Together with Delhi, also in India, Calcutta is one of the world's seven cities with the worst air pollution. This is caused by cars and trucks belching out smoke and fumes into the air.

Key points

- The population of Calcutta is growing at a fast rate.
- High population density is causing problems such as overcrowding, pollution and the lack of clean water.

Hong Kong: High Population Density

Key words

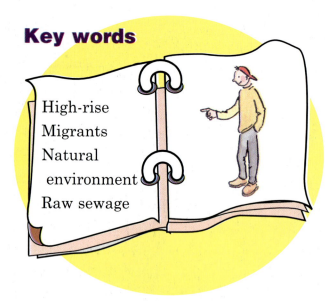

High-rise
Migrants
Natural
environment
Raw sewage

Hong Kong, in east Asia, is like Calcutta, a very overcrowded place. Unlike Calcutta, Hong Kong is in a developed area.

Hong Kong belongs to China. It was ruled by Britain from the 1840s but was handed back to China on 1 July 1997.

It is a very modern city of **high-rise** apartment blocks, many factories, modern shops and businesses. All of this is cramped into a very small space.

Area: 1,092 km²
Population: 7,211,000 (in 2001)
Density: 6,603 persons per km²
              (compare Ireland
              which has a population
              density of 56 persons
              per km²)

Hong Kong includes four main areas:
- Hong Kong Island, on the south side of the Victoria Harbour, the main business area.
- Kowloon, the peninsula on the north side of the harbour.
- The New Territories, between Kowloon and the Chinese border. These make up 90% of Hong Kong's land area, and 30% of the population.
- 260 outlying islands.

High population density has created a number of problems:
- Overcrowding
- Lack of open spaces
- Pollution.

Overcrowding

Huge numbers of **migrants**, mainly Chinese, have flooded into Kowloon and onto the island of Hong Kong. There is great demand for housing in this very small space. As a result, homes are small and rents are high. The streets between the apartments are narrow and dark. Unable to get houses, some people live on houseboats in the harbour.

Lack of open spaces

Although it is a wealthy area, there is a lack of open spaces where people can go to relax. Every square metre is needed for buildings. Even graves in a cemetery had to be stacked up on concrete terraces.

Hong Kong has spread up the slopes of surrounding hills. The rock and soil have been dug away to level out the land for houses and streets. This rock and soil has been brought to the sea where it is used to make new land. The **natural environment** is being destroyed.

Pollution

With the pressure of so many people in such a small space there is a problem getting rid of waste from the city. As in Calcutta, rubbish piles up.

Rubbish piles up in Hong Kong

Noise pollution is a serious problem in Hong Kong. The causes include:
- Aeroplanes constantly landing and taking off.
- Heavy traffic on the roads.

Air pollution is caused by:
- Heavy local traffic.
- Gases from factory chimneys.

Water pollution is caused by:
- The pumping of **raw sewage** into the sea.
- Waste from factories, in particular, heavy metals from steel factories.

Key points

- Hong Kong is an area with high population density in the developed (rich) world.
- High population density in Hong Kong is causing problems of overcrowding, lack of open space and serious pollution.

Mali: Low Population Density

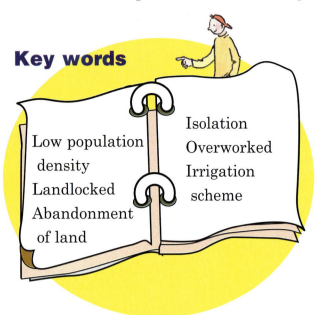

Key words

Low population density
Landlocked
Abandonment of land

Isolation
Overworked
Irrigation scheme

Your study of Calcutta and Hong Kong showed you that there are problems when many people crowd into a small space. Let's look at a case where having too few people causes problems. Mali is one of the world's poorest countries. Mali is a country in West Africa with **low population density**.

Mali lies at 17°N of the equator. It is a **landlocked** country i.e. no part of the country is beside the sea. It is 15 times the size of Ireland but only has a population of 11 million. This means it has a low population density of 9 people per km². This has caused many problems for the people of Mali.

The problems caused by low population density are:
- **Abandonment of land**
- **Isolation**
- Shortage of clean water.

B *Landlocked Mali*

Abandonment of land

In Mali 65% of the land is desert or semi-desert. This percentage is increasing. So few people live in these isolated areas that the government has not spent money on bringing in water. People have had to abandon (leave) the land in parts of Mali.

This is happening because the land that is available for farming is being **overworked**. Land has to be given time to rest in between harvests. It needs fertilisers and watering. If this doesn't happen, the soil breaks down, dries out and is blown away. It is then abandoned.

'Don't know when I'll be back again'

The problem could be lessened if there was an **irrigation scheme**. Water could be brought to the farms by pipes and poured onto the land. Then crops could grow. This has not happened because of low population density.

Isolation from markets

Mali is a landlocked country – it has no coastline. So, as it cannot send its goods to markets by ship, it must use roads to transport its goods.

It is costly to build roads to isolated areas. The government has not put money into road links because the people are so spread out across the land. Of the 15,000 km of highway only 1,827 km is paved. Without an improvement in the road network, Mali will stay isolated from markets. This problem is caused by Mali's low population density.

Lack of clean water

Much of the country of Mali lies on the edge of or within the Sahara Desert. Water is scarce. The problem of supplying water to remote areas is similar to the problem of building roads. It is too costly to bring a water supply to so few people. Nearly half of the 11 million people of Mali do not have safe water to drink. It is only in settlements of higher density such as Bamako, the capital, that you find safe water schemes.

A village pump brings hope

Often people drink water from polluted sources, such as old wells and ponds. Deaths from diseases caused by unclean water are common, especially among children under one year old.

Key points

- Mali in West Africa is one of the world's poorest countries.
- Mali is a country with **low population density**.
- Low population density leads to **abandonment of land**, **isolation** from markets and poor services.

The West of Ireland: Low Population Density

Key words

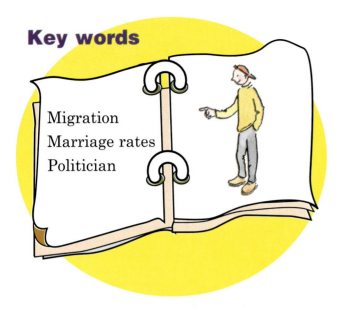

Migration
Marriage rates
Politician

This is above the national average of 23.3%. Because there was a shortage of women, **marriage rates** were lower. In other words, the number of people marrying dropped. Birth rates, in turn, dropped. Population densities dropped further.

'I have only my dreams' Ⓑ

The West of Ireland is an area of low population density. This has caused many problems. People are moving from areas of low density into areas of high density.

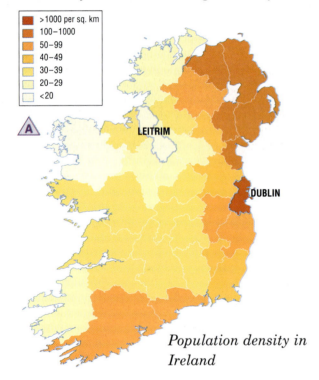

	>1000 per sq. km
	100–1000
	50–99
	40–49
	30–39
	20–29
	<20

Ⓐ

LEITRIM

DUBLIN

Population density in Ireland

Low marriage rates

Migration has always been part of life in the West of Ireland. In the past farms were usually left to sons. Young women often had to move away for work. An older, male population was left behind. In counties such as Leitrim, over 36% of the male population is aged over 65 and single.

Abandonment of land

When young people move away, older people are left behind. They often find it difficult to look after their farms well. If land is not looked after, it will become run down. Without improvements, soil becomes infertile. It will be hard to make a living from it. These old people will move away too. The farm will be abandoned.

Once this fall in population happens, smaller villages or towns lose services such as the post office, bus service, bank or the Garda station. As you learned in your study of Mali, you need a certain number of people to support a service to make it worth spending money on it. For the people who are left behind this drop in services is difficult for them. It may push them to leave too. There is then a cycle of population loss.

This pie chart shows that, in Ireland, more and more people are living in cities and towns.

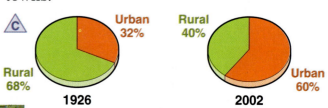

Ⓒ

Urban 32%

Rural 68%

1926

Rural 40%

Urban 60%

2002

Political and economic isolation

The Dáil

Politicians are elected to the Dáil to look after the needs of the people who elect them. The Dublin region, with more than one-third of Ireland's population, has many politicians representing it.

The West of Ireland has a lower population density and so this region, although a larger area than the east, will have fewer politicians acting for it.

Look at chart E. It shows the changes in the percentages of elected politicians from each province. The number representing Leinster (the east) is rising steadily while the number representing Connacht (the west) is falling.

It is hard to attract businesses to areas where there is low population density. This is because of poorer transport links with the main market areas. This is true for the West of Ireland. There has not been the same level of spending on roads in the west as there has been in the east. This is directly related to the issue of low population density in the West of Ireland.

When population falls services are cut back

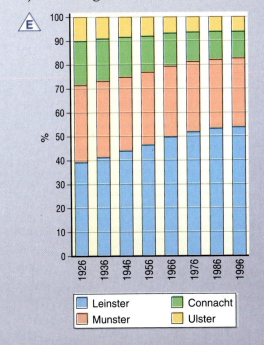

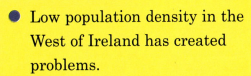

Key points

- Low population density in the West of Ireland has created problems.
- Land has been abandoned.
- **Marriage rates** are low.
- Services become poor in areas of low density.

REVISION EXERCISES

Write the answers in your copybook.

1 Hong Kong lies in:
- Southern Asia
- The Middle East
- Eastern Asia
- South America

2 Bustees are:
- Cattle farms in Mali
- Tropical red soils
- Volcanic material
- Slums in developing countries

3 Shanty towns are usually found in:
- Big cities in developed countries
- New towns like Shannon and Tallaght
- Big cities in developing countries
- Rural villages in developing countries

4 An area noted for its high population density is:
- Mali
- West of Ireland
- Hong Kong
- Amazon Rain Forest

5 Some of the problems associated with over-population are:

(a) Overcrowding   (d) High infant mortality rates

(b) Low marriage rates   (e) High living standards

(c) Pollution

Not all the answers are correct. The **three** correct statements are numbered:

(i) a, b, c   (iii) a, b, d

(ii) a, c, d   (iv) b, c, e

6 In your copybook match each letter in column X with the number of its pair in column Y.

X	Y	ANSWER
A High population density	1 West of Ireland	A = _____
B Low population density	2 Calcutta	B = _____
C Low marriage rates	3 Mali	C = _____
D Bustees	4 Hong Kong	D = _____

7 The city of Calcutta has areas of unplanned housing called bustees. Imagine that you are visiting one of these areas. Write a letter to a friend describing what you see.

8 Areas of high population in the developed world have many problems. Name **two** of these problems and in the case of **one** of those problems clearly describe how it affects the quality of life for the people living there.

9 Large areas of Mali suffer from drought. How has this affected the lives of the people? In your answer refer to availability of clean drinking water and water for crops.

10 Describe **two** problems that develop when a country has a low population density.

11 Imagine you are living in the West of Ireland. Write a letter to your local TD asking him/her to highlight one of the main problems in your area during a meeting of the Dáil. Outline your solution to this problem.

12 Copy the extract below and fill in the gaps.
The West of Ireland is an area of _____ population density. In counties like Leitrim this has led to a _____ marriage rate and a feeling of isolation. Over ___ % of the population are males over the age of 65. Many of the _____ people have left for the larger cities where work is available. There are _____ TDs representing this area in the Dáil. Often this leads to the area being neglected when the government is giving money.

5 Migration – The Movement Of People

Key words

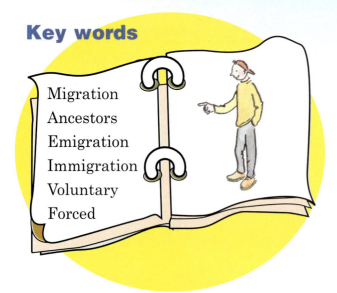

Migration
Ancestors
Emigration
Immigration
Voluntary
Forced

Migration means to move from one place to settle in another. It is a word we use to describe the movement of animals to warmer places in the winter. They migrate in search of food.

Migration of birds in winter

Humans have also migrated across continents in search of food, shelter, safety and good weather. Migration is still very much part of life! With improvements in transport, migration is even easier. It is very likely that you too will migrate to another country to study, work or simply for adventure!

Ireland – A History of Migration

Migration is part of Irish history. Our **ancestors** migrated from Europe to settle in Ireland many years ago. In the middle of the nineteenth century, during the Great Famine, thousands of Irish people migrated to England and the United States.

Up until the 1970s the movement of people was mainly out of the country. This was called **emigration**. Since then, movement is both into as well as out of the country. Movement into the country is called **im**migration. Movement out of the country is called **em**igration.

Between April 2001 and April 2003 over 137,000 **immigrants** came to Ireland – 35% of them were Irish people returning home. Other immigrants came from countries in Africa, Asia, Eastern Europe and other places.

Contestants in the Rose of Tralee competition are likely to say 'My ancestors came from Ireland'

Immigration and change

This pattern of immigration is bringing great changes to Ireland. Among the changes is an increase in the numbers of people in the country.

Just as Australia, the USA and England had to plan for the thousands of Irish migrants who travelled abroad in the past, Ireland now has to plan for the many immigrants who are coming to this country in search of a better life.

Some things to consider will be:

- The extra people in the labour force, many of whose skills are important to the country.
- The extra cost of providing services such as health, education and housing.
- The changes in the cultural make-up of the country as different races and nationalities settle down together.

All of this needs careful planning by the government.

Types of migration – voluntary or forced movements

In many cases, people migrate because they want to. It is a **voluntary** movement. They decide for personal reasons to migrate to another place.

In other cases, people might not have any choice. They may be **forced** to move because it is a matter of life or death. They might be forced to flee because of famine, war or religious persecution in their home country.

Types of migration – organised or individual

If the migration is part of a planned or organised event, this is an organised movement.

If a person is not part of a large organised group of migrants, this is an individual movement.

Key points

- People migrate out of choice or because they are forced to.
- A person can migrate as part of an organised group or on his/her own.
- Governments need to plan for **migration**.

Migration – Push and Pull Factors

Key words

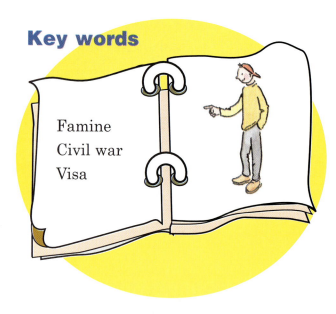

Famine
Civil war
Visa

In the previous section you learned that the migration of people from one area to another is a natural part of modern life. Nowadays, it is easier than ever to move! Transport is better now. We have television and the Internet to spread information about other places.

In 1965 the total number of migrants worldwide was 75 million. It has grown hugely since then and has now passed 120 million.

Why people move

The many reasons people leave their homes for a new one can be divided into push and pull factors.

- Push factors are the reasons that push a person out of their homelands.
- Pull factors are the reasons that 'pull' or attract people to come to an area.

The women from the Philippines who work as nurses in Irish hospitals were drawn or 'pulled' here by the higher pay and the work permits offered to them.

When the potato crop failed, causing the Great Famine of the nineteenth century in Ireland, people were pushed or forced to emigrate to England or America.

Pull factors
- Good jobs and higher pay
- Cheap land to farm
- The promise of work permits
- Better education

Push factors
- Widespread unemployment
- Lack of farmland
- **Famine**
- **Civil war**

TO A BETTER WAY OF LIFE

Barriers to migration

People may want to move, but it is not always possible. Things that stop people from moving are called barriers. Let's look at these barriers.

Visa requirements

In many cases a government puts a limit on the number of foreigners they allow to settle in a country. A person needs permission, a **visa** or work permit, to live or work in another country. Some people risk entering countries without permission. But if they are discovered entering illegally they will usually be deported (sent back). The requirement of a visa is therefore a barrier to movement.

Poverty

The cost of migration can be very expensive. For many poor people it is an option they cannot even consider. A lack of money therefore is a barrier to migration.

Emotional ties

For many people migration means that they may never see their homeland again. The ties of friends and family may be too strong. The pain of separation may be too much for people. These emotional ties are barriers to migration.

Conchita's story

It is not unusual for parents in my small town in the Philippines to migrate in order to earn money for their children's education. My husband Pepe is still waiting for a work permit. Hopefully it will arrive soon. We will work very hard for a few years and then we hope to return to our country and build our own house. I miss my family. Leaving them was the hardest decision I can remember ever having to make.

Time passes slowly when you are away from your loved ones. The only thing that keeps me going is the airline ticket on the mantelpiece. It reminds me that soon it will be holiday time and I will be with my family again.

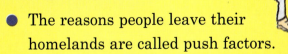

Key points

- The reasons people leave their homelands are called push factors.
- The reasons they are attracted to an area are called pull factors.
- Most people migrate for work reasons.
- Barriers stop people from moving.

A Voluntary Movement – From the West to the East of Ireland

Key words

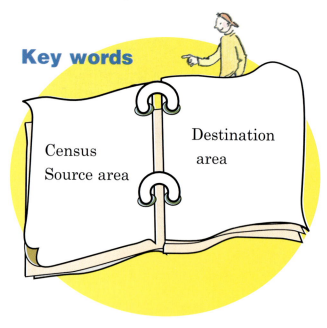

Census
Source area

Destination
area

People may move out of choice. When this happens it is called a voluntary movement. The movement of people from the west to the east of Ireland is an example of a voluntary movement. It is a free choice.

A

Abandoned village in the West of Ireland

There is a history of movement towards Dublin from the West of Ireland. Since the population was first counted in the **census** of 1841, the population of Co. Leitrim in the west has fallen steadily. This is due to migration out of Leitrim.

The population in and around Dublin has continued to rise as people move from rural areas towards the city.

Changes in Population			
	1841	1951	2002
Leitrim	155,297	41,209	25,799
Dublin	372,773	693,022	1,122,821

People mainly move to Dublin as they are pulled there by the promise of jobs and education. Government offices, the head offices of many companies, hotels and the many entertainment theatres are examples of the opportunities for jobs in Dublin.

Family members join each other as time passes. Social reasons, like nightlife, then begin to attract people into the city region.

Many people move away from rural areas like Leitrim. They are pushed to move because there are few opportunities for work or further education.

This has effects in both the **source area** (Leitrim) and in the **destination area** (Dublin). The source area is the place *from* which people move. The destination area is the place *to* which people move.

Effect of migration

The effect on the source area

When people move away from the West of Ireland:

- Population numbers fall.
- A high proportion of old people are left behind.
- The birth rate falls when young people leave.
- Services such as schools and hospitals are cut back.
- Investment in industry falls. Businesses are unwilling to set up where there are a small number of young workers.

The effect on the destination area

When people move into the Dublin area:

- The population rises in the city and the surrounding counties of Wicklow, Kildare and Meath.
- New houses are built. Schools are provided. The number of hospitals increases. Shopping centres are built.
- More transport links are added – DART line, Luas, M50 and the Port Tunnel.
- Industries are attracted to the region.
- House prices rise, so people move out towards the suburbs where land is cheaper.

People have to travel long distances to work in the city. Traffic congestion increases.

But it is not all rosy in the city.

Key points

- People migrate from the west coast to the east coast of Ireland.
- This brings changes to the **source area** – the West of Ireland.
- It also brings changes to the **destination area** – the Dublin area.

An Organised Migration – The Plantation of Ireland

Key words

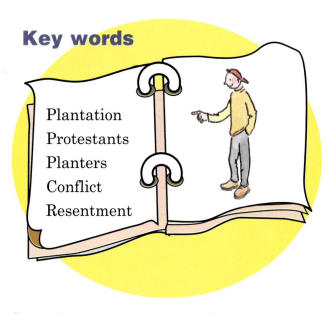

Plantation
Protestants
Planters
Conflict
Resentment

Sometimes a government plans or organises a movement of a group of people to a new place. This is called a **plantation**. This happened in Ireland during the sixteenth and seventeenth centuries. During what became known as the Plantation of Ireland, the native Irish were forced off the land to make way for these new settlers. One area that was settled at this time was Ulster. Let's look at the Ulster Plantation in more detail.

The Plantation of Ulster

King James I of England brought the Scottish **Protestants** to Ulster. This migration, believed to include well over 100,000 Scottish Protestants, mainly took place between 1607 and 1697. These new settlers came to six of the nine counties of Ulster.

Counties Armagh, Cavan, Derry, Donegal, Fermanagh and Tyrone were settled by planters

The new settlers were pulled by the promise of cheap land. They were given holdings of between 400 and 800 hectares, land that had been owned and farmed by the Gaelic-speaking native Irish. This migration had effects on Ulster which have lasted to the present day.

Effects of the Plantation of Ulster

The **planters** built new towns. Their style of town was new on the landscape. In the centre was a square or a triangle called a diamond. The most important buildings, the town hall and the courthouse, were built here. Tradesmen such as carpenters lived in these towns.

They also built villages in every county. These villages had a cluster of houses. The most prominent building in the village was the castle where the local court was held. The castle also provided a place of refuge in times of attack from the native Irish. Other buildings in the village included the mill and the malt house. So, there were active small-scale industries. Not everybody was a farmer!

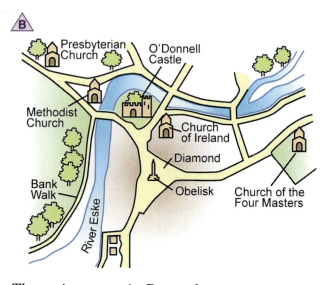

The main square in Donegal

When the planters arrived from England and Scotland much of the land was forested. Our native trees (deciduous), which provide hard wood, started to disappear from the landscape. Deciduous trees are those such as oak and ash that lose their leaves in winter. The planters cut down many of these forests to make way for farmland. Today, few of our forests grow deciduous trees.

The planters carefully planned how they could make the best use of the land. They divided it into large estates of up to 2,000 acres. Many fenced off their land. They organised market areas where goods were bought and sold and advice was traded. They also had knowledge of and access to the bigger English market where they could sell their goods.

The settlers introduced a new culture into Ireland. Before the **plantations** the Gaelic language was the language of Ireland.

With the arrival of the settlers, the English language was introduced. The settlers also introduced the **Protestant** religion to Ireland. This replaced the Catholic religion in many places.

The present day **conflict** between the two cultures in Ulster can be traced back to the time of the plantations. The push of the native Catholic population to the poorer lands caused **resentment** to grow between the two groups. Over the course of 200 years the divisions deepened between the two cultural groups – the Protestant Loyalists and the Catholic Nationalists.

Key points

- The **Plantation** of Ulster was an organised migration.
- It mainly took place during the seventeenth century (1607–97).
- The **Plantation** of Ulster brought such lasting changes as planned towns, the English language and a history of **conflict**.

International Migration

Key words

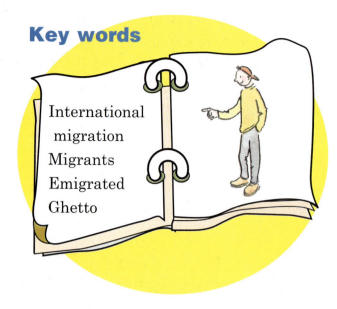

International
migration
Migrants
Emigrated
Ghetto

International migration is the movement of people from one country to another. International migration is now at an all-time high. About 145 million people worldwide lived outside their native countries in the mid-1990s, and the number is increasing by 2 million to 4 million each year. Many **migrants** today are from the poorer countries of the world.

Irish migration to the USA

For a long time people from Ireland have migrated to the USA. They have gone in search of jobs and a new life. You have already learned that huge numbers of Irish people **emigrated** to America during and after the Great Famine of the nineteenth century.

B	Irish Migrants to USA	
Years		**Total Number**
1840–1850		2 million
1820–1920		5 million

Many migrants stayed with their own nationalities, setting up ghettos in the large cities of the USA. A **ghetto** is an area that is associated with a particular group. These early migrants were poor and could not afford decent housing so the ghettos were often overcrowded. Many of them lived in areas such as the Bronx in New York.

Destinations

Main movements of migrants seeking work

World movement of people

People chose to move. They believed that they would find better opportunities for work in the USA. At first, these Irish labourers worked mainly in the building industry. They started off working at the bottom of the ladder in unskilled jobs.

Soon, they or their children, moved into higher paid jobs like the police and fire services.

Every year we are reminded of this migration when we see the many St Patrick's Day parades held in cities around the USA.

St Patrick's Day parade in New York

The Irish culture was very strong in many cities so others were encouraged to follow. They organised social events and classes that would bind the Irish community together, reducing feelings of loss.

Irish culture is alive and kicking in the USA

Soon they built schools for their children. Once these were set up migrants wrote back home to bishops asking for an Irish priest to join them. Then churches were built. They became important social meeting places. Notre Dame University in Indiana grew from this background.

E	Percentage of Irish Emigrants to the USA	
Years		**% of Total Emigrants**
1987		24.6
1997		14.1
2001		11.6

Fewer Irish emigrants are choosing to go to the USA

In more recent times Irish people have moved to other international places such as France or Spain. For many, who are now well educated, language is no longer a barrier.

Key points

- **International migration** is more common today than in the past.
- Most **migrants** are from the poorer areas of the world.
- Over 2 million Irish people left for the USA during the famine of the 1840s.
- Migrants were pulled by the promise of jobs.

An Organised Migration – South America

Key words

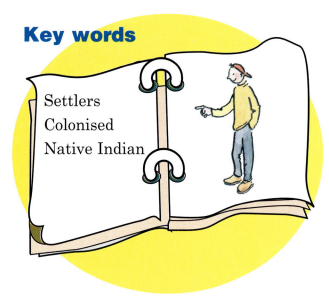

Settlers
Colonised
Native Indian

A	Changes in the Number of Native Indians in South America	
Year	**Population**	
1500	15 million	
1900	3.5 million	
2000	0.2 million	

Look at the map. It shows the colonisation of South America.

European **settlers** found their way to South America during the fifteenth and sixteenth centuries. They came in search of wealth and the promise of new lands. The settlers mainly came from Spain and Portugal. They **colonised** (took control) of the countries of South America. These colonies later became independent countries.

Many **Native Indians** were killed or driven into the remote parts of the forests.

Chart **A** shows the changes in the numbers of Native Indians in South America as migration took hold.

COLOMBIA was colonised by the Spanish in 1510

ECUADOR was colonised by the Spanish in 1532

BRAZIL was colonised by the Portuguese in 1500

PERU was colonised by the Spanish in 1531

CHILE was colonised by the Spanish in 1541

ARGENTINA was colonised by the Spanish in 1580

Portuguese spoken
Spanish spoken

B

Colonisation of South America

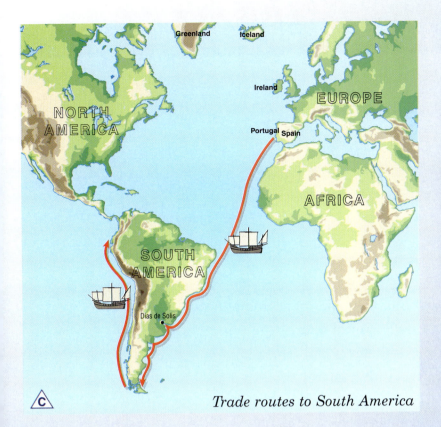

Trade routes to South America

- Many of the **Native Indians** are still poor. They often have bad health and poor access to healthcare.

A favela

The results of this migration are still obvious today.

- Spanish and Portuguese are the main languages spoken in South America.

- These **Native Indian** people are powerless. They are not treated fairly. They can be pushed off their land by big companies who want their land for timber or to rear cattle.

- Most of the main cities are on the coast. This is where the colonisers first settled and from which they traded.

Indians lost their lands to big foreign companies

Key points

- South America was mainly **colonised** by the Portuguese and the Spanish.
- Portuguese and Spanish became the main languages in South America.
- The Native Indians are still poor and powerless.
- As a result of this migration the main cities in South America today are on the coast.

The coastal city of Rio de Janeiro, Brazil

REVISION EXERCISES

Write the answers in your copybook.

1 The movement of people between home and work in towns and cities every day is called:
   ● Population density
   ● Emigration
   ● Immigration
   ● Commuting

2 Circle the correct answer in **each** of the pairs of statements below.
   (a) Lack of work in an area is a **pull/push** factor.
   (b) A good social life is a **pull/push** factor.
   (c) We refer to those who leave Ireland in search of work as **emigrants/immigrants**.

3 Match each letter in column X with the number of its pair in column Y.

X	Y	ANSWER
A International migration	1 Ulster	A =
B Organised migration	2 Jobs	B =
C Push factor	3 Family left behind	C =
D Pull factor	4 Ireland to the USA	D =
E Barrier to movement	5 War	E =

4 Which of the following is called an organised migration?
   ● The colonisation of South America
   ● From the west to the east of Ireland
   ● From the city centre to the suburbs
   ● To Britain from Ireland

5 Colonisation exists when:
   ● Two countries agree to trade with each other
   ● One country controls other countries
   ● One country gives food aid to another
   ● Two countries agree to a ceasefire

6   Explain the following words:
   ● Immigration
   ● Emigration
   ● Planters
   ● Voluntary
   ● Colonisation

7   The Plantation of Ulster and the European colonisation of South
    America are examples of organised migration.
    Choose one of the above migrations and explain **three** effects the
    migration had on the area to which the people moved.

8   (a) Name the four counties which have
       had a decrease in population and
       the three counties which have had
       the greatest increase in population
       from 1991–96.
    (b) In most of the counties with big
       increases in population the greatest
       growth has been in or near large
       towns and cities. Give **three**
       reasons why people migrate to large
       towns and cities.

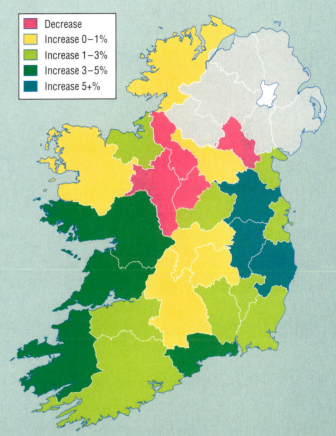

Decrease
Increase 0–1%
Increase 1–3%
Increase 3–5%
Increase 5+%

9   Describe **one** example of migration of
    people you have studied. In your
    answer refer to the push and/or pull
    factors involved.

10 Describe **three** ways that out-migration has affected rural
    (country) areas like the West of Ireland.

11 Give **two** reasons why people are pulled to the Dublin region.
    Explain how **each** of these reasons has affected those already
    living in the Dublin region.

6 Settlement of Ireland

Key words

Site
Mode
of transport
Monuments

Mesolithic
Lumberjacks

Archaeologists believe that the first human settlements in Ireland were made about 6000 BC. A settlement is a place where people set up a home. The place where a settlement was first built is called the **site**.

Where people settle is usually connected to:

- Where they have migrated from.
- Where they can get to easily using their **mode of transport**.
- Where there is food and water.
- Where they will be safe and able to defend themselves if needed.

When you travel around Ireland you will see a wealth of **monuments** and other evidence showing the different migrations of people into this small country. The sites for these settlements were chosen carefully. You can understand why they settled in each location using the points above.

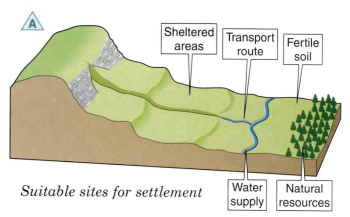

Suitable sites for settlement

Sheltered areas · Transport route · Fertile soil · Water supply · Natural resources

Mesolithic people

Ireland's first settlers were Middle Stone Age (**Mesolithic**) hunter-gatherers who used stone implements (tools). They travelled in boats and settled in coastal areas. They settled where there was water, fish and wild animals. They needed to be near the coast as there were no roads. Also the land beyond the coast was covered in thick forests. There were no **lumberjacks** then! No mechanical saws! They travelled by boats along seas and rivers. They left behind little heaps of waste in which traces of their food and pottery were found.

Early settlers chose a riverside location

Neolithic people

The next group of settlers were farmers. They arrived in the Neolithic era (about 3000 BC). They settled where the soil was light and could be worked by their stone tools. They grew crops and tamed and raised animals. These conditions are found in the Burren, Co. Clare.

The Neolithic people chose sites beside rivers. This provided a water supply and a link to the sea for trading. These first farmers liked to settle in higher places because they could be easily defended.

Starting about 2000 BC, they built the massive stone tombs called 'megaliths'. These are shown in red on an Ordnance Survey map. The most famous of these is at Newgrange, Co. Meath.

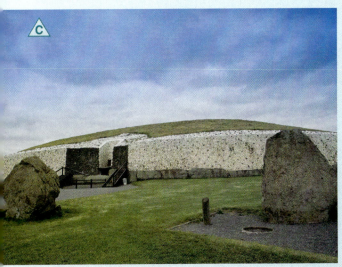

Megalithic tomb at Newgrange

The Celts

The Celts started off in central Europe, but at an early date they spread into southern France and northern Spain. A small group arrived in Ireland in about 600 BC. A major migration of Celts arrived in Ireland about 350 BC, and it is certain that the Celts had settlements throughout the island by 150 BC. They brought new skills with them. They had mastered the use of iron.

Our Celtic ancestors had fair skin, red-blond hair, and were taller and larger than other people of the time. Traces of their settlements are found in the many ring forts dotted around the country. Ring forts were often built on hills or hilltops. The fact that they built forts would mean that they needed to protect themselves from attack from earlier settled groups. They settled near water supplies so they could fish and travel easily.

Evidence of the Celtic settlement of a ring fort, is seen in placenames such as lis, dun, caher and rath.

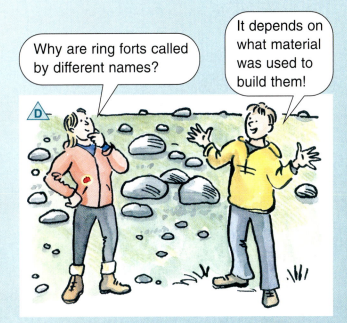

Key points

- A settlement is a place where people set up homes.
- The first human settlements in Ireland date from about 6000 BC.
- In the pre-Christian era groups such as the **Mesolithic**, Neolithic and Celtic settled in Ireland.

The Viking Settlements

Key words

Viking raids
Monasteries
Plunder
Settlers

The era of Viking settlement began in AD 795 when a raid took place on Lambay Island off the coast of Co. Dublin. Further **raids** followed. The Vikings attacked and **plundered** the monasteries. They stole treasures from the monasteries as the monks were not good at fighting and were unable to protect them.

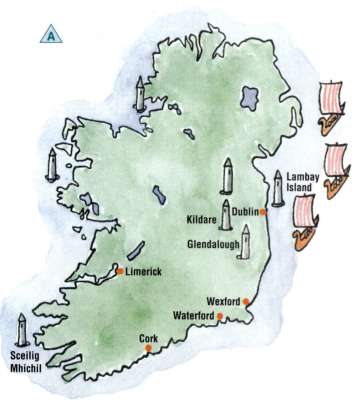

This map of Ireland shows the monastic settlements which were raided by the Vikings as well as showing Viking settlements

As the raids increased, the Vikings began to **plunder** settlements further inland.

The Vikings liked Ireland. Winters were milder than in Denmark and Norway so it was easier to rear animals here. They decided to settle down in Ireland.

The Vikings set up towns along the coast in places such as Dublin, Waterford, Wexford, Cork and Limerick.

These coastal sites were popular because:
- They could dock their longboats safely in a sheltered harbour.
- They could set sail again to trade all over Europe and the Middle East.
- There was a plentiful supply of water for drinking and cooking.
- Fish was a staple food. They ate salmon and trout from rivers; herring and mackerel, caught with iron hooks from the sea; oysters, cockles, mussels and scallops from the seashore.

They often chose to set up their settlements at the bend of a river or on high ground. Those sites would be easy to defend from attacks from the native Gaelic tribes.

Dublin – a Viking settlement

Dublin is an example of a Viking settlement. It was built on a site that could be easily defended, a ridge of high ground overlooking the River Liffey. The **settlers** surrounded it with a bank made of earth to protect it from attack. The Vikings docked their longboats in a sheltered pool called Dubh Linn (Black Pool).

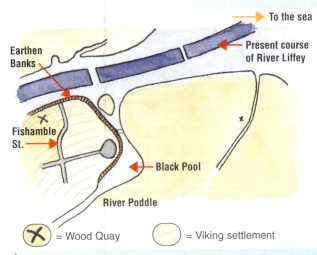

B *The Viking settlement of Dublin*

Earthen Banks

To the sea

Present course of River Liffey

Fishamble St.

Black Pool

River Poddle

✗ = Wood Quay   = Viking settlement

Coins, combs and weighing scales have been found on excavated sites such as Wood Quay in the heart of medieval Dublin. These items show the variety and importance of trade in medieval Dublin. Dublin was well planned in terms of where the various tradespeople worked and lived. The street names and the street map give us this evidence.

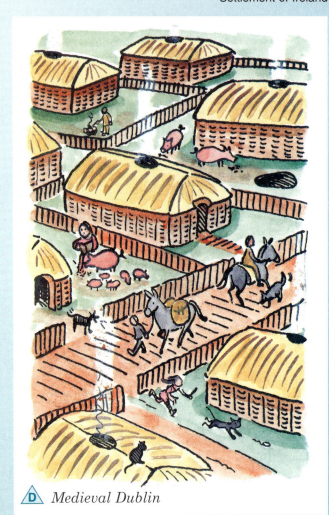

D *Medieval Dublin*

Essex Quay has also been excavated

Fishamble Street follows the same route as it did in medieval times. This is evident from the early maps of the city.

Houses were dark, damp and cold in winter and hot, smelly and poorly ventilated (aired) in summer. Fire was a risk as the fire was built in the centre of the house. Inside the house, light was poor. It usually came from rushlights made from mutton fat.

Trade was an important activity in Viking settlements. The Vikings set up towns along the coast so that traders would have access to them.

Key points

- The Vikings came from Norway and Denmark.
- They set up towns along the coast on sites that could be easily defended, and from which they could trade.

The Norman and the Plantation Settlements

Key words

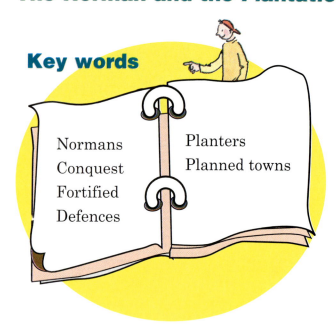

Normans
Conquest
Fortified
Defences

Planters
Planned towns

A

The Norman settlement era began in the twelfth century. **The Normans** came from Normandy in Northern France. They settled in Britain first and later invaded Ireland. These invaders were called Anglo-Normans.

The Anglo-Norman **conquest** of Ireland began in 1169. A force of Norman soldiers sailed over from Wales, and captured Wexford and Waterford. A year later they took Dublin, and then, over the next 80 years, they moved out in all directions, until they held about 75% of the island.

Everywhere they went in Ireland, the Normans built castles and towns. They were the first settlers in Ireland to build inland towns i.e. towns not on the coast.

They settled near water supplies. They **fortified** (made strong) the **defences** started by the Vikings. They built stone walls with look-out towers and strong city gates to keep out attackers. St Audoen's Gate in Dublin, still visible, dates from this period. One such gate, St James's Gate, is famously associated with Guinness!

The plantations

The next major period of settlement in Ireland was during the plantations of the seventeenth century. A plantation is a planned migration of new settlers to an area. The landscape in Ireland today shows clear evidence of this type of planned settlement. There were three main plantations in Ireland. These took place in:

- Laois and Offaly
- Munster
- Ulster.

You have already studied the Ulster Plantation in the chapter on migration.

B

Planned town of Birr, Co. Offaly

The **planters** settled on large farms where the land was fertile and crops would grow well. Local people would work on these farms or rent some land. The landlords often travelled outside Ireland bringing back new ideas about farming.

The planters settled on the lowlands where it was easy to build road links to the coast for trading. They exported their crops to England.

They chose areas where there was space to plan new settlements.

The planters left their mark on the Irish landscape. They built large houses on big estates surrounded by native trees, such as beech and oak, many of which are still seen today.

They introduced town planning to Ireland. Streets were widened and housing and other services were carefully planned. Grand Georgian houses, made of brick, were built around the centre. These houses had rounded arched doorways with fanlights. Trees lined the streets.

The many **planned towns** around Ireland date from the plantations. These were, and still are, very attractive towns. A planned town has:

- A big open area at the centre. It may be a square or a triangle in shape. It is often called a diamond.
- Important buildings at the centre. These include the town hall, the courthouse and the jail.
- Streets which are set at right angles to the square.
- Two-storey houses which are built in terraces (in a row) along these streets.
- A Protestant church placed at the centre.

Remember, for the Junior Certificate exam, you are expected to become an expert on one of the settlement eras described in this and the last section.

Planned towns in Ireland

Planned Towns	County
Kenmare	Co. Kerry
Sneem	Co. Kerry
Adare	Co. Limerick
Abbeyleix	Co. Laois
Birr	Co. Offaly
Enniskerry	Co. Wicklow
Donegal	Co. Donegal

Key points

- The Anglo-Normans set up towns along the coast and inland. They defended their towns with castles and stone fortifications.
- The **planters**, who came from England and Scotland, were given large estate farms. They built **planned towns**.
- Evidence of these settlement areas can be found in buildings and placenames.

REVISION EXERCISES

Write the answers in your copybook.

1 In Dublin and other areas, large brick houses with wrought iron railings and rounded arched doorways were built during which era?
   ● Viking
   ● Neolithic
   ● Norman
   ● Plantation

2 Explain **each** of the following terms:
   ● A megalith
   ● A defensive site
   ● A dun

3 People have always chosen particular sites on which to settle. In the case of **one** settlement that you have studied discuss **two** of the reasons why that site was chosen.

4 Write the correct answers for **each** of the statements below:
   (a) The Neolithic people were **farmers/factory workers**.
   (b) The Vikings came from the **Mediterranean area/Scandinavia**.
   (c) The Normans came from **Spain/France**.

5 In your copybook match each letter in column X with the number of its pair in column Y.

X	Y	ANSWER
A The early settlers left these	1 Castles	A =
B These settlers built the first towns	2 Megaliths	B =
C The Normans left these monuments	3 Planned town	C =
D Birr, Co. Offaly is one of these	4 Vikings	D =

6 Why did the Vikings settle in Ireland? Write a short account of **one** Viking town that you have studied. In your answer refer to the types of activities carried out in the town and how we now know what they worked at.

7　What kind of sites did the Normans choose for settlement? Give details of **two** factors influencing their choice of site.

8　Describe what a Norman settlement might have looked like. Refer to the site chosen and the evidence from the monuments left behind.

9　Why did the planters come to Ireland? How did they change town planning in Ireland?

10　Look at the drawing of a planned settlement. Name three features that show that this is a planned settlement.

11　Look closely at the map.
(a) Name two areas of Ireland where there is little evidence of Anglo-Norman settlements.
(b) Name three towns that lie on the same river, all dating from the Norman period of settlement in Ireland.
(c) Name three Viking seaports. Name one feature that they have in common.

7 The Distribution of Settlement in Ireland

Key words

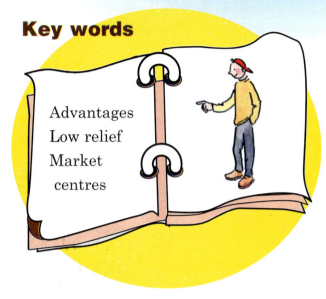

Advantages
Low relief
Market centres

Towns grow in places where there are natural **advantages**.

Low relief

Most Irish settlements are found in areas of **low relief**. These will be lowland areas where slopes are not too steep or the land too high.

Lowlands are suitable for settlements such as towns and cities because it is easier and cheaper to build houses and roads on more level land. It is also easier to provide services like water and electricity to settlements on lowlands rather than highlands.

Example: Newbridge, Co. Kildare

Quality of the land

If the soil around a town or village is fertile and easily worked, towns can develop as **market** and manufacturing **centres**. What happens is as follows:

- Local farmers produce food crops, such as wheat or barley, or they rear animals, such as cattle or pigs.
- The farmers travel to the local town to trade their produce.
- People gather in the town for the market.
- Mills might be set up to make flour from the wheat.
- A creamery or co-op might be built to process the milk into cheese or butter or dairy spread.
- A factory might be set up to make sausages or rashers.
- So, a town grows because it was a market centre in the middle of an area of fertile soils.

Example: Nenagh, Co. Tipperary

The opposite is also true. Where there is infertile soil and poor farming, it is unlikely that market centres will develop.

Lowlands attract settlement

Rivers

Towns and cities tend to be built near rivers or where two rivers meet (confluence). This is the pattern all over Ireland.

Examples: Cork on the River Lee; Galway on the River Corrib

Resources

A resource is something that is useful. Many towns grew where they had natural resources like minerals, beautiful scenery or fish nearby.

The nearby lead and zinc mines have been important to the development of the town of Navan in Co. Meath. People found work in the mines so houses were built. Shops and schools were set up for the families of workers and so the town developed.

Another example is the resource of beautiful scenery around the town of Killarney. The natural beauty of the surroundings has led to an important tourist industry. Hotels and bed and breakfasts have been set up. Restaurants and pubs do a good trade and so the town has developed.

The town of Killybegs has grown because of the natural resource of fish. This town is the most important fishing port in Ireland. A variety of jobs such as boat building, net repair and fish processing are hugely important to the people who live in Killybegs.

The government has given grants to people to encourage them to stay in the area.

The resources of the sea attract settlement

Think of your own town. Why did it develop? Is it an historical town? Is it on lowland? Does a river flow through it? Is its development connected with a natural resource?

Key points

- Towns are set up in areas where there are natural **advantages**.
- These advantages include **low relief**, good soil, access to rivers and natural resources.

Patterns that Settlement Make in Our Landscape

Key words

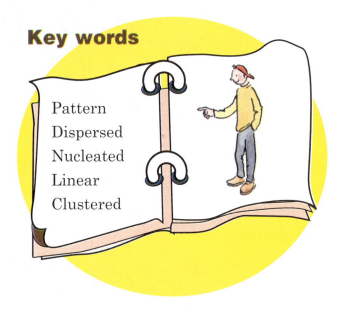

Pattern
Dispersed
Nucleated
Linear
Clustered

Geographers are interested in the way settlements are spread out (distributed) across the landscape. Once people settle, it is possible to see certain patterns in the spread of their houses, villages, towns and cities.

To understand the idea of **pattern** look at the picture which shows you how individual houses can be part of a pattern.

The pattern of houses can be **dispersed**, **nucleated** or **linear**.

Dispersed pattern of settlement

Houses are spaced out across an area. **Dispersed** could also be described as a scattered pattern. This usually means a farmhouse on its own piece of land, separated from its neighbours. Farmers like to live on their own plot of land with easy access to their farms.

Nucleated pattern of settlement

Houses stand together in a group. The buildings in villages, towns and cities form a **nucleated** pattern. In rural areas, farmhouses clustered or grouped together form a nucleated pattern.

The **nucleated** pattern allows people to share services. It is cheaper to supply a group of settlements with a service, e.g. a water supply, than to supply settlements that are spread out.

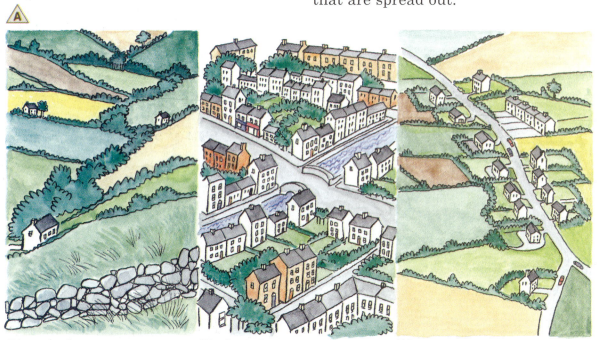

Dispersed   *Nucleated*   *Linear*

Linear pattern of settlement

Houses are built in a line. This happens where houses are built alongside a road, or on the coast facing the sea or by a river.

Farmers sell sites for new houses with access to the main road. This eases the supply of services to people. Houses in a **linear** pattern along roads can access the bus service more easily than if they were scattered in a more dispersed pattern.

Cities and towns form patterns on the landscape

In the same way as individual houses form a pattern with other houses, larger settlements such as villages, towns and cities can also form a pattern together. If you look at the map of Ireland, you will see that Irish towns and cities form dispersed, nucleated or linear patterns.

● **Dispersed**: This pattern is one where towns are scattered across the landscape. Many of the planned towns in Laois and Offaly form a dispersed pattern.

Dispersed pattern of towns

● **Clustered**: This pattern is one where towns are clustered around a major city. Many towns on the edge of Dublin or Belfast form this pattern.

Cluster of towns around Dublin

● **Linear**: Where towns are spaced out along a line you find a linear pattern. This would be common along a major road or along the coastline. The road from Dublin to Cork for example is lined with towns along the route.

Towns form a linear pattern along roads

Key points

- Towns form a **pattern** on the landscape.
- Towns may form a **dispersed** pattern.
- Towns may form a **linear** pattern.
- Towns may form a **clustered** pattern.

The Spread of Towns in Ireland

Key words

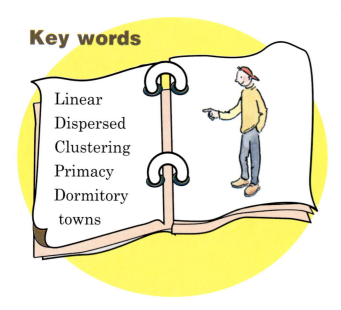

Linear
Dispersed
Clustering
Primacy
Dormitory
towns

The origin of linear settlement is in access to the sea

Now that you know what the patterns look like, let's discover the reasons for these patterns.

Social and historical reasons

Many Irish towns form a **linear** pattern around the coastline.

Knowing the history of Irish towns helps us to understand why many are dotted around the coastline in this way. When the Vikings, and later the Normans, built settlements, they needed to get water and food close to their settlement. They wanted a good place to dock their ships and a site they could defend easily in times of attack. They wanted to trade in foreign places. For all these reasons, the pattern of setting up at the mouth of rivers around the coastline was adopted. This explains why many of Ireland's towns form a linear pattern around the coast.

Dispersed towns or villages

Dispersed towns

A further piece of history explains the **dispersed** or spread out pattern of towns in the Midland counties of Laois and Offaly. Towns such as Abbeyleix, Mountmellick and Birr are examples of planned towns. They were built by landlords to be market and service centres for what they produced on their large estates. These towns are spaced out in a dispersed pattern linked to the large estates of the planters.

Clustered towns

A more recent pattern to develop in Ireland is the **clustering** of towns around major cities. This has happened around the fringe of the Greater Dublin Area. Look at map Ⓓ. This pattern is a result of government planning.

The growth of clustered settlements

This pattern has come about because of the **primacy** of Dublin.

The primacy of Dublin

The word **primacy** means that one city in a country is dominant. Dublin, for example, is ten times bigger than Cork, the next largest city in Ireland.

Dublin has become a primate city because jobs and services have been centred in Dublin. People now commute (travel) on a daily basis into the city centre for work or services. The cost of houses in Dublin has forced many people out to the outlying towns to live where houses are cheaper.

They live out past the edge of the city in what are known as **dormitory towns**. They are so called because it would seem that many people only sleep in these towns. Travel and work take up most of their days.

If you were an astronaut in space looking down on Ireland you would see the patterns that towns make on the landscape:

- Dispersed
- Linear
- Clustered.

Key points

- Towns form a **linear** pattern around the coastline of Ireland.
- Planned towns in the Midlands are spread out in a **dispersed** pattern.
- Towns form a **clustered** pattern around Dublin.
- Dublin is a primate city.

REVISION EXERCISES

Write the answers in your copybook.

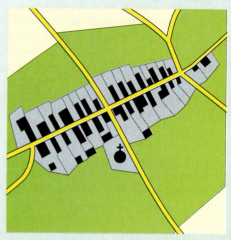

1 What is the settlement pattern shown on the sketch?
- Clustered
- Dispersed
- Nucleated
- Linear

2 Match each letter in column X with the number of its pair in column Y.

X	Y	ANSWER
A Houses are scattered	1 People who travel to work each day	A =
B Houses are in clusters	2 Dispersed pattern	B =
C Houses are in a line	3 Nucleated	C =
D Commuters	4 Linear	D =

3 Towns develop and grow where there are favourable conditions like lowland, fertile soils, rivers and natural resources. In the case of a town that you have studied, examine the importance of **two** of these conditions for its growth.

4 Explain the following words:
- Infertile soil
- Confluence of rivers
- Natural resource

5 Killybegs, Co. Donegal, is Ireland's major fishing port. Explain **two** reasons why it has grown over the years.

6 With the help of a diagram show the differences between the **three** types of settlement patterns: nucleated, dispersed and linear.

7 In the case of **one** of the patterns listed in question 6, explain why this pattern of houses developed in our landscape.

8   In recent times a number of towns have grown up around Dublin.
    Name **one** of these towns and suggest **two** reasons why it has
    increased its population.

9   Explain the following terms:
    ● Primacy
    ● Dormitory town
    ● Commuters

10 Dublin is a primate city. Explain what this means.

11 Look at the picture showing a settlement pattern.
    (a) Name the pattern of settlement shown in the cartoon.
    (b) Explain two reasons for this settlement pattern.

8 New Settlement Patterns

Key words

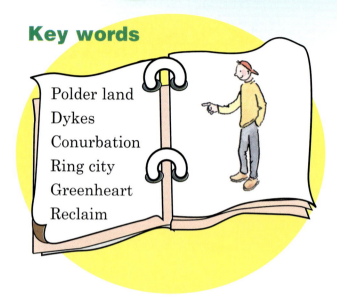

Polder land
Dykes
Conurbation
Ring city
Greenheart
Reclaim

Planned settlements

The population of the world is growing very fast, so more land is needed for uses such as houses, factories and roads.

This has led to the planning of new settlements. A new settlement is a 'planned' space. Like a painter in a studio, planners are often working with a blank canvas. They choose a suitable space on the map and decide on the best way to use it. They consider modern needs such as bicycle lanes or football stadia and include them in their plans.

Let's look at an example of planned settlements that were built on land that was reclaimed from the sea. This happened in the Netherlands, also called Holland.

Case Study: The Netherlands

Population:	About 16 million
Size:	About half that of Ireland
Main cities:	Amsterdam, Rotterdam, Utrecht and The Hague
New land:	The Zuyder Zee Project

The Netherlands is a flat, low-lying country, criss-crossed by waterways (canals and rivers). There are many canals in the Netherlands because nearly a quarter of the country lies below sea level.

Below sea level

Land that used to be below the sea is called **polder land**. Water has to be artificially drained off and carried by canals into the sea. Otherwise, the land would be flooded.

This land has to be protected by high banks called **dykes**.

In the Netherlands water is pumped off the land and drained into canals

Almost the entire west and north of the country is polder land. The land of Holland reaches over 200 metres above the sea level only in the south-west of the country.

Population distribution in the Netherlands

The population in the Netherlands is not evenly spread (distributed). Close to 45% of the population lives in the area where the cities of Amsterdam, Rotterdam, The Hague and Utrecht are located.

These cities and their urban fringes have spread out so much that they almost meet. When cities melt together in this way, we call it a **conurbation**. Notice the word 'urban' in this word. Urban means town or city.

In the Netherlands a group of cities has 'merged' together to form one huge urban area called 'The Randstad'. It is as if Dublin and Galway each spread out so much that they met.

Randstad means **ring city**, so called because it is a circular shape. It has a space in the centre called a **greenheart**. This greenheart is used for farming and recreation, hence its title 'green'.

The sprawling of these cities into each other was causing huge problems for the Dutch. The government needed to deal with overcrowding, traffic congestion, water supply and space for recreation. They had to think about where to build new settlements for the growing population.

The problem was that they just didn't have space in their country for new settlements.

They decided to **reclaim** land from the sea. Reclaiming land means draining the water out of the sea, drying it out and providing new land. The Netherlands is no stranger to this kind of work.

New planned settlements were built on new space in what was once part of the sea.

Key points

- New settlements are needed as the world gets more crowded.
- 'The Randstad' is a **conurbation** in south-west Holland where cities and towns have merged.
- To take the pressure off 'The Randstad' new settlements were planned and built on land **reclaimed** from the sea.

The Zuyder Zee Project – Polder Land for New Settlements

Key words

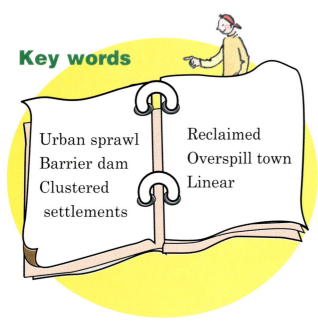

Urban sprawl
Barrier dam
Clustered
settlements

Reclaimed
Overspill town
Linear

These reclamation works started in 1920. The plan included:

- A **barrier dam** to cut off the inlet of the Zuyder Zee from the sea.
- Five new polders covering 1,650 square kilometres of land.
- An inland lake called Lake Ijssel or the Ijsselmeer.

The Dutch created new land for planned settlements. They reclaimed land from a large sea inlet in the north of the country, called the Zuyder Zee project.

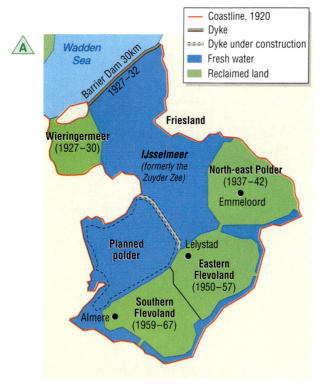

A section of a barrier dam

The aims of the project were:

- To create new land for farms.
- To create new settlements to lessen the problem of **urban sprawl** in the Randstad.
- To create a new freshwater lake for water supply and for recreation.

New settlements on the polders

The new settlements on the polders were carefully planned. The largest town, situated in the middle of the new area, is an important service centre. It provides schools, hospitals and large department stores.

Smaller centres such as villages were also planned. They form a pattern of **clustered settlements**. They serve a smaller area with shops selling products needed on a daily basis such as milk and bread.

Planned settlements on the North-east Polder

Almere – a city built on the polders

The city of Almere is one of these planned new towns. It is built on the **reclaimed** land of the southern Flevoland polder.

It is called an **overspill town** because it took some of the population from Amsterdam.

Almere is at the centre of a large area with many roads linking it to smaller towns and villages. It serves the people in the area around it by having a variety of shops, schools, churches, hospitals and places of entertainment. New houses were built here. Road and rail links connected Almere and Amsterdam. Open spaces were left for parks and leisure activities.

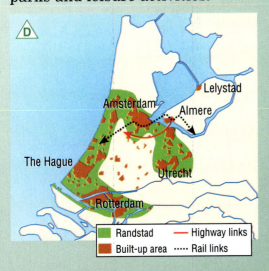

Land was set aside for industries so that jobs could be created in the area. In this way, people did not have to travel to Amsterdam.

The planning of farms

As this was a new plan, planners were able to draw up the best possible layout for settlements and farms.

On polders used for farming, farmhouses are built in groups of three or four along roads. This allows people to be near each other for security and to share the cost of services like electricity. This pattern of settlement along the line of the road is called **linear**.

The fields are laid out and surrounded by canals. The canals carry all the excess water away and drain it into the large lake of Ijsselmeer. As most of this country is below sea level, the land will always have to be drained. This is done using diesel pumps.

Key points

- The Zuyder Zee project was a project to **reclaim** new land from the sea.
- This newly **reclaimed** land was used for planned settlements.
- In planned settlements, large towns are at the centre of a network of smaller villages around it.
- Farmhouses in the **reclaimed** areas of the Netherlands are built in a line along roads.

REVISION EXERCISES

Write the answers in your copybook.

1 Areas of land reclaimed from the sea are called:
- Polders
- Favelas
- Colonies
- Lagoons

2 A planned new town in the Netherlands is:
- Amsterdam
- Rotterdam
- The Hague
- Almere

3 In your copybook write the correct answer for each of the statements below.
   (a) The river that flows through the Netherlands into the North Sea is the **Rhine/Seine**.
   (b) The barrier dam created the freshwater lake called the **Zuyder Zee/Ijsselmeer**.
   (c) A town built to take pressure off a larger town is called an **overspill town/conurbation**.

4 Match each letter in column X with the number of its pair in column Y.

X	Y	ANSWER
A Greenheart	1 Area of open space	A =
B Dyke	2 Cities that melt into each other	B =
C Urban sprawl	3 A bank to stop flooding	C =
D Conurbation	4 The spread of towns and cities	D =

5 From which sea was land in the Netherlands reclaimed? Why was there such a demand for more land by the Dutch?

6 What is a polder? Describe **one** way in which **settlement** and **land use** on polders are different from other areas.

7   It is rare for many town planners to plan a town from green fields or reclaimed land. Discuss **two** of the advantages that this brings to the project. (Hint: transport routes)

8   Describe **one** disadvantage of living on a polder.

9   Draw a carefully labelled diagram showing how farms are laid out on polder land (include roads that are very straight).

10 Look at the map.
    (a) Name the cities marked A–D.
    (b) Name the features marked E and F.
    (c) Explain what the word **conurbation** means.
    (d) Describe the area called Randstad.

Randstad

11 Name a new town in the Netherlands that you have studied.
    (a) Draw a sketch to show the location of this town.
    (b) Explain why this town is called an 'overspill' town.

12 Look at the following pie charts.
    (a) What percentage of the land of the Netherlands is in the Randstad?
    (b) What percentage of the total population of the Netherlands lives in the Randstad?
    (c) Explain one problem that arises from high population density in the Randstad area.

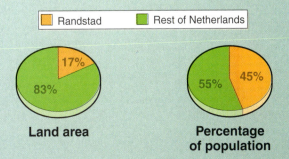

Randstad   Rest of Netherlands

17%
83%
Land area

45%
55%
Percentage of population

Functions of Nucleated Settlements

Key words

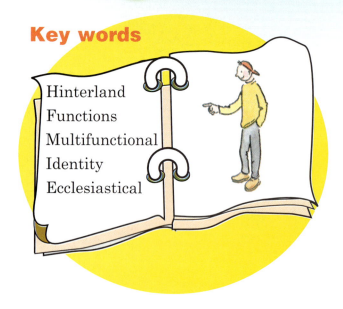

Hinterland
Functions
Multifunctional
Identity
Ecclesiastical

What is the purpose of a town? Think of your own local town. What services does it offer you?

You may go to town to buy clothes. The cinema and the Garda Station may be in the town.

Obviously there are a number of services offered in your local town or village. The town or village serves the surrounding area (the **hinterland**). These services are also called **functions**.

As towns have many **functions**, we can say that towns are **multifunctional** settlements.

Sometimes when you look at a town, one function or activity may stand out more than any other. This function may give the town an **identity**. Look at the following examples.

- A *village* settlement is small. Basic services are found, such as shops, a pub and a post office e.g. Kilflynn Village, Co. Kerry.

- A *market* settlement is one where the main function is trading. There may be a large market square to show this e.g. Nenagh, Co. Tipperary.

- A *defence* settlement is one where the main function in the past was defence. You may still see a castle or part of the old wall of the town e.g. Kilkenny City.

- A *resource based* settlement is a town that grew because of one particular resource. A mining town would have a mine nearby e.g. Navan, Co. Meath.

Multifunctional settlement

- A *port* settlement is a town that is noted for its shipping function e.g. Rosslare, Co. Wexford is a ferry port.
- A *residential* settlement is one where a town's main function is housing, e.g. Leixlip, Co. Kildare.
- A *religious* or **ecclesiastical** settlement is one that is noted for religious functions. You would see a cathedral, monastery or abbey e.g. Armagh City.
- A *recreational* settlement is one which is mainly noted for entertainment or holiday activities e.g. Bundoran, Co. Donegal.

Let's take a closer look at three of these examples.

Armagh – a religious settlement

Armagh City is an important **ecclesiastical** (religious) centre. It is the headquarters of both the Catholic and Protestant Churches in Ireland. Armagh has two cathedrals. Today, Armagh has other functions including housing, shopping and leisure activities.

Protestant cathedral in Armagh City

Kilkenny – a defence settlement

Kilkenny, built on the River Nore, developed around a Norman castle. A large stone castle was built because the river formed a natural moat, which meant the castle could be easily defended if attacked.

Kilkenny Castle has been beautifully restored and is an important tourist attraction.

As Kilkenny grew, other functions such as trade and industry developed. The city is well-known for its arts and crafts.

Kilkenny Castle

Navan – a resource based settlement

Navan first developed because it was at the meeting point of two rivers, the Boyne and the Blackwater. It was an important crossing point for travellers and traders.

When lead and zinc were found in nearby mines, the town grew in importance. As more jobs became available people moved to the area. Services such as housing, banks and shops developed. Navan is now a large multifunctional settlement.

Lead and zinc mines in Navan

Key points

- **Functions** have to do with the purpose of a town.
- A settlement can get its **identity** from a particular function.
- All towns are **multifunctional**.

Settlement in an Irish River Basin

Key words

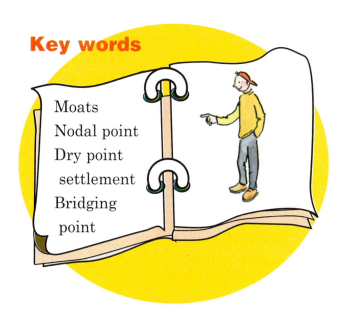

Moats
Nodal point
Dry point
  settlement
Bridging
  point

A

Let's look at the main functions of three settlements in the basin of the River Shannon.

The River Shannon is the longest river in Ireland. The source of the river is in the Cuilcagh Mountains of Cavan and it flows into the sea beyond Limerick. Settlements were set up along the Shannon because:

- The river provided a supply of water.
- The land around it was good for farming because of fertile soil.
- Some places along the river could be defended well in times of attack e.g. bends were natural **moats**.
- Crossing points on the river were good places to trade.

No wonder many settlements grew up along its banks!

B

Limerick

Limerick – a port settlement

- *Port function*: Limerick is an important port dating back to the Vikings. It is built on the long sheltered estuary of the Shannon. The many routeways that meet at this busy crossing point carry traffic to and from the port.

- *Retail function*: From its earliest days, it was a centre for business. Today, its many shops make it a magnet for shoppers.

- *Education function*: Limerick is a major centre for education. The University of Limerick is located there.

Carrick-on-Shannon – a recreational centre

- *Market function*: Carrick-on-Shannon is an important business and market centre for the surrounding area.
- *Transport function*: The town is a **nodal point** (centre of routeways), a busy crossing point on the Shannon. Roads such as the N4 and the railway link it to Dublin and Sligo. The opening of the canal at Ballyconnell, Co. Cavan now brings more movement between Ballyshannon, Co. Donegal and Limerick City.
- *Recreational function*: People have always enjoyed boating and fishing on the river. Many boating companies have set up businesses here. The tourist activity has resulted in the growth in night-time entertainment.

Carrick-on-Shannon

Athlone

River Shannon

Athlone – a defence settlement

- *Defence function*: The Normans built a castle at Athlone. The site chosen was on high ground so flooding was not a threat. This is called a **dry point settlement**. Later, an army barracks was built in the town linking back to its original defence function.
- *Transport function*: Athlone is at a **bridging point** on the River Shannon. The ancient routeway between Dublin and Galway crossed the river at Athlone. It is still a busy crossing point.
- *Market function*: Athlone is the largest Midland town in Ireland. Due to its central location, it serves a wide hinterland with goods and services.

Residential function: With the growth in the number of industries and services, the built up area of Limerick has spread out into the countryside.

...ook at a map of Ireland and find the ...iver Shannon.

Key points

- The Shannon River has attracted settlement along its banks.
- Carrick-on-Shannon, Athlone and Limerick are important towns along the Shannon.
- They serve a number of functions.

Settlement in the Rhine River Basin

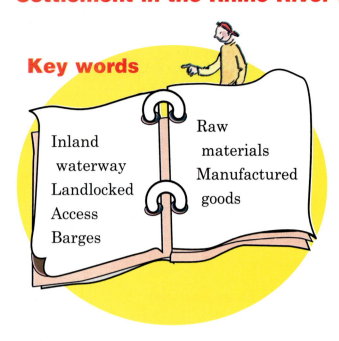

Key words

Inland waterway
Landlocked
Access
Barges

Raw materials
Manufactured goods

People have always liked to settle near rivers. This is because:

● The fertile alluvial soils of floodplains mean that crops can be grown successfully.

● The fast flow of water near the source of the river can be tapped for hydroelectricity. This can then be used in homes and factories.

● Rivers are great transport routes especially for heavy and bulky goods.

River Rhine

The River Rhine is the focus of many large settlements. It is the most important **inland waterway** in Europe. It is about 1,200 km from its source in the Swiss Alps to its mouth at Rotterdam in the Netherlands. Many towns along its banks have grown because of the trade along the river.

The Rhine has given **landlocked** countries like Switzerland **access** to the sea. The river has linked most of central Europe by water, easing the way by which goods can be transported, bought and sold.

The River Rhine has many tributaries such as the Moselle and Ruhr which flow into its main path. Over time, many canals have been built to link these rivers. In this way, large areas of central Europe have become connected.

The River Rhine

Constant traffic brings people

The Rhine is one of the busiest rivers in the world. Ships and **barges** carry **raw materials** and **manufactured goods** along this waterway. Many of the cities along its course can now handle huge barges. Factories have also set up all along its banks. These industries use the river for transport as well as a source of water.

Wherever there are industries there are jobs. Wherever there are jobs people settle. When people settle they need services such as shops, schools and hospitals. In this way, towns like Basle, Duisburg and Rotterdam have developed.

Let's look at Basle, Duisburg and Rotterdam in more detail.

Basle

- Basle is an inland port. It is Switzerland's main port. Roads lead from the port to many areas of central Europe.
- Industry is an important function of Basle. The large supply of water and the good transport system have encouraged the growth of industries. Chemical companies and engineering firms have set up here.
- Education is now an important function of Basle. The many colleges and universities offer courses related to its industrial and service functions.

Duisburg

- Duisburg is the largest inland port in Europe. Therefore, transport is a major function of the city. Duisburg is built where the rivers Rhine and Ruhr meet.
- Duisburg serves a huge industrial hinterland called 'the Ruhr'. Factories producing iron, steel and chemicals have set up here because there are plentiful supplies of water and coal. There is also an excellent river transport system.
- This densely populated city now has an important residential function.

Rotterdam

- Rotterdam is the largest port in the world, so trade is an important function. It has excellent deep-water facilities, which means it can handle huge oil tankers from the Middle East.
- It is the centre of a vast commercial area. Many big firms such as banks, insurance companies and department stores have set up here. The city has a large migrant population who find jobs in these services.
- As part of the densely populated area of the Randstad, Rotterdam has an important residential function.

Key points

- The River Rhine is Europe's most important waterway.
- Rotterdam is the world's busiest port.
- Duisburg is the world's busiest inland port.
- The Rhine gives Switzerland a link with the sea at the inland port of Basle.

Change of Functions over Time

Key words

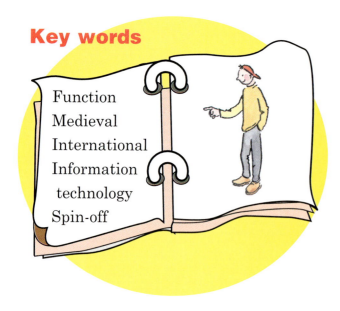

Function
Medieval
International
Information
technology
Spin-off

Settlements grow for a reason. This is called a **function**. In some cases, this original function is no longer important. Changes have happened.

Let's look at the changes in the functions of two settlements.

Kilkenny

Kilkenny, on the River Nore, grew around a Norman castle. It was the site of many parliaments during **medieval** times. The

city no longer functions as a defensive settlement. In recent times, large-scale industrial development has come to Kilkenny.

Kilkenny is at the heart of a rich farming area. Farming has provided many raw materials that can be used in factories. One such raw material is barley. This is used in the making of Smithwicks beer.

Brewing

The Smithwicks Brewery is in the heart of Kilkenny. It is part of the Diageo group. The factory brews 1.2 million hectolitres of, mainly, Budweiser and Kilkenny Irish Beer. This is sold in areas such as the Far East and New Zealand.

Food processing

Glanbia plc is an **international** food company. The company's headquarters is located in Kilkenny. It sells dairy and meat products across the world.

The company employs over 7,500 staff worldwide, many of whom are in Kilkenny.

Glanbia ranks as one of Europe's largest dairy companies, and one of the world's largest cheese manufacturers.

Craft and design

Pottery being made at Kilkenny Design, Kilkenny

Since the government set up Kilkenny Design Workshops in 1965, Kilkenny has become famous for crafts. Artists from all over the world work here. Kilkenny Design has contributed greatly to the development of Kilkenny as an important centre for industry.

Information technology

At present Kilkenny is trying to attract companies involved with **information technology**. These companies have been drawn to Kilkenny because of the modern, purpose-built business parks and other services.

So, Kilkenny has changed from a defence settlement built around the castle to an important industrial centre.

Navan

Navan developed as a market centre because it is located at the meeting point of two rivers. It is a nodal point. In recent years mining has become important.

Tara Mines

The largest lead and zinc mines in Europe were discovered in Navan in 1973. Jobs became available in the mines and in **spin-off** industries. There was a boom in the local building industry. New housing estates were built. Roads were improved. The mines brought more money to the town as workers spent their wages in the local area. Services such as shops, garages and entertainment grew.

Navan – a nodal point and mining town

Key points

- The functions of settlements can change over time.
- As time goes on a town will develop a number of functions.
- Kilkenny changed to an important centre for large scale industrial development.
- Navan's functions changed as mining became important.

REVISION EXERCISES

Write the answers in your copybook.

1 A historic wall around a town or city shows that which of the following functions was important in the past?
   - Residential
   - Defence
   - Transport
   - Recreation

2 The statement which best describes the course of the River Rhine is:
   - Rises in the Netherlands and enters the sea from Switzerland
   - Rises in Germany and enters the sea from Switzerland
   - Rises in Switzerland and enters the sea from Belgium
   - Rises in Switzerland and enters the sea from the Netherlands

3 Which of the following towns have developed in the Rhine River basin?
   - Munich, Berlin, Freiburg
   - Dresden, Leipzig, Halle
   - Basle, Duisburg, Rotterdam
   - Hamburg, Hannover, Rostock

4 The original function of Kilkenny was:
   - Recreational
   - Religious
   - Educational
   - Defensive

5 A nodal point is:
   - Where many routeways meet
   - The site of many industries in a town
   - A headland that juts out
   - Functions that change over time

6 Explain the following:
- A multifunctional town
- A resource based settlement
- A recreational settlement

7 Write the correct answer for **each** of the statements below in your copybook.
(a) The River Shannon rises in the **Cuilcagh Mts/Wicklow Mts**.
(b) It ends its course near the city of **Galway/Limerick**.
(c) One large town along its route is **Mullingar/Athlone**.

8 Explain what is meant by the functions of a town. In your answer name **one** town and give at least **one** function that town offers.

9 Name **three** towns that developed along the banks of the River Shannon. In the case of **one** of these towns describe the many functions that it now provides for local people.

10 Draw a map showing the course of the River Rhine. Write a brief account of its importance as a routeway in Europe.

11 In the case of **one** town or city that has grown up near the banks of the **River Rhine**, describe **three** functions that this town offers to the people who live nearby.

12 Match each letter in column X with the number of its pair in column Y.

X	Y	ANSWER
A Brewery	1 Service	A =
B Kilkenny Design	2 Smithwicks	B =
C Function	3 Lead and zinc	C =
D Navan	4 Arts and crafts	D =

13 Explain **one** reason why the functions of a settlement may change over time. Include an example in your answer.

Key words

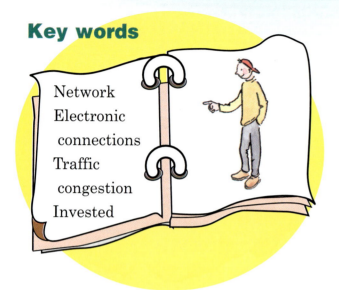

Network
Electronic connections
Traffic congestion
Invested

A **network** of road, rail, airways and waterways links people and places. Telephone and **electronic connections** also make links. These links help settlements like cities and towns to grow.

Communication links

Road links

Roads are the most important way of transporting people and goods around the country.

Ireland has 100,000 km of roads. If you stretched all the roads out together in a line they would wrap around the middle of the world more than twice.

Rise in car ownership

In Irish society over the last ten years there has been a big increase in the number of cars.

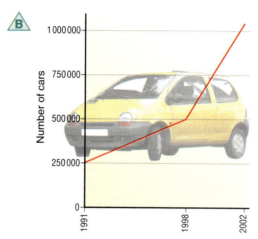

Increase in car ownership in Ireland, 1991-2002

The rise in the number of vehicles has led to severe **traffic congestion** on our roads. This congestion occurs along the main roads leading into large towns. It is at its worst on roads to and from cities during peak hour traffic.

The Irish road network is being improved to cope with this increase.

Who says travel is fun?

Investment in roads

The National Roads Authority (NRA) was set up in 1993. Their job is to make sure that there is a safe and efficient network of roads in the country.

With the help of the EU, the government has **invested** (spent) in providing a good road network. 50 billion euros will be spent on roads between 2000–06.

Major improvements have been planned to link the towns of the poorer regions of the west and north of Ireland to the better off regions of the east. These poorer regions will become better off as the road links improve. Look at map ⒠.

Types of roads

There are different types of roads.

- *National primary roads* link the main ports, airports, cities and large towns. Look at map ⒠ showing these roads. Notice how they link Dublin to the cities of Cork, Limerick, Galway, Belfast and Waterford. Many of these roads have dual carriageways and others have sections of motorway.

- *National secondary roads* may link smaller towns and cities. These roads link the national primary roads to each other.

- *Regional roads* connect smaller settlements to the national primary and national secondary roads.

- 'Other' less important roads link smaller settlements.

Important road projects

In recent years, huge sums of money have been spent on the following projects:

- *M50 motorway*: This is a ring road to the west of Dublin. It has given rise to new settlements close to this routeway.

- *By-passes*: These roads avoid the towns by running along the outside rather than through them. Many towns have already been by-passed e.g. Athlone, Drogheda and Kildare. They are spared the traffic jams that used to clog up the main streets. This leads to the growth of these towns.

Key points

- Road transport is the most important means of transport in Ireland.
- There has been a huge investment in improving the network of roads throughout the country.
- Good road links will aid the development of poorer regions.
- Transport plays an important role in the growth of settlements.

Settlements and the US Interstate Highway System

Key words

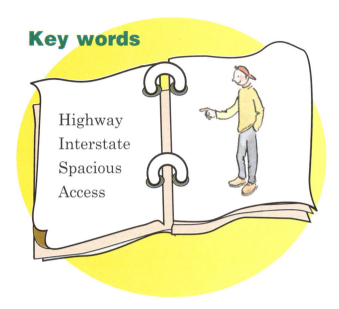

Highway
Interstate
Spacious
Access

Let's look at the road network in the USA. Americans call their main roads **highways** or freeways. A network of **highways** criss-crosses the USA from New York state in the east to the state of California in the west.

It is called an **interstate system** because it links up the states of America from coast to coast.

Fifty or so years ago the only links between the cities of the states were simple two-lane roads. Most of them had many stop signs, traffic lights, railway and cattle crossings. They were too slow for a growing economy.

The **interstate highway system** replaced these roads. It is based on the German autobahn system. It was built for economic and social reasons.

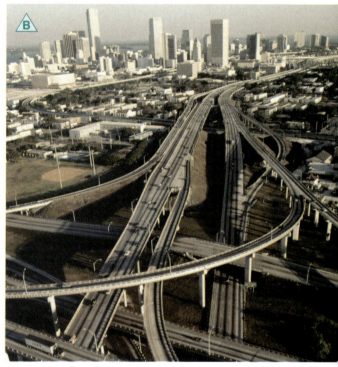

The US highway system

Towns along the US interstate highway system

Some of the **highways** pass through vast empty areas. You could travel for long distances and never meet a settlement. These roads link the big cities and make it easy to move people and goods from state to state. In other cases settlements have grown along the routeways.

Influence of the highways system on settlement

Work

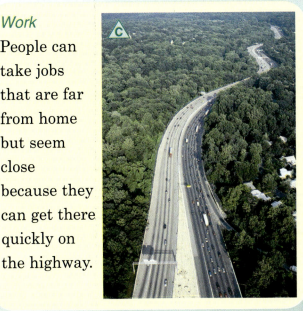

People can take jobs that are far from home but seem close because they can get there quickly on the highway.

Housing

People can enjoy living in **spacious** houses way out in the suburbs. The **interstate highway system** has aided the growth of settlements beyond the main cities.

Shopping

People can drive out onto the highway and reach shopping malls and recreation centres with great ease.

Leisure

The **interstate highway system** means that people can take long trips at weekends to the mountains or the coast. This has led to thriving businesses in motels, restaurant stops and service stations.

Healthcare

People have **access** to big hospitals and specialist care. They can move quickly over great distances to get the healthcare they need.

Key points

- The greater movement of people today has led to a fast and efficient road network such as the **interstate highway system**.
- The **interstate highway system** has influenced the development of settlements on the landscape.

Settlements and the French Railway System

Key words

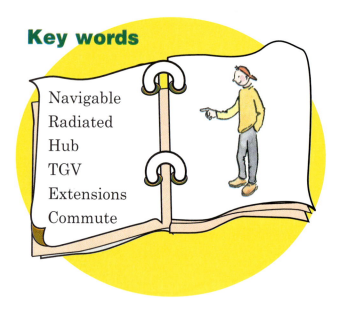

Navigable
Radiated
Hub
TGV
Extensions
Commute

Settlements grow if they are linked to other settlements by an efficient transport system. Let's look at the French railway network and see how it has helped the development of towns and cities in France.

History of the French railways

The main French railway system was built after 1842. It was built to transport goods in and out of factories.

Before railways became important, bulky goods were transported along rivers or canals. Many factories were built close to the coalfields because coal was burned to produce power to run machines. The coal and the goods from the factories were moved along **navigable** rivers and canals.

However, because many of the growing towns and cities were not linked by water, it was decided to build railways.

From the beginning, the French government built long stretches of railway lines in eastern France along the German border. This was to defend that border, not to link settlements. The first completed lines, however, **radiated** out of Paris. This connected France's major cities to the capital.

The government advised that every town with a population of around 1,500 inhabitants should have a rail link.

Paris is still the **hub** (centre) of the railway system in France.

France's TGV (high speed trains)

The French railway system today has around 40,000 km of track. Its most modern and admired train is the **TGV**.

TGV is France's *train à grande vitesse*, which means 'high-speed train'. The project was started in the 1960s. It connects Paris to other cities in France and to some other neighbouring countries, such as the United Kingdom and Belgium.

There are planned **extensions** of the high-speed lines to other major cities in Europe.

This fast train system was introduced to compete with air travel. The first public TGV run was made from Paris to Lyon. The public found it so comfortable and speedy that they stopped using the air route between these two cities.

Tourists travel to the Riviera. This builds stronger links between cities like Paris and Marseilles. In this way development of the South of France has been speeded up.

In time, an extensive railway network will link most European cities. Then it may be possible to choose more freely where to live. More settlements may develop in areas not previously considered.

These examples show how the development of settlements is closely linked to the development of communication links such as the railways. People like to settle in places that are well connected and linked to other places.

The TGV speeds through the countryside

TGV

Possible speed	515.3 km/h
	(320.2 mph)
Usual speed	300 km/h
	(185 mph)
No. of TGV train sets	360

Impact of the TGV

Workers in towns such as Lille **commute** to Paris daily. This encourages development of other regions beyond Paris. It now takes only 90 minutes to travel between Brussels and Paris.

Key points

- The French railway system was originally built to transport goods from factories.
- The **TGV** is the French high-speed train.
- Paris is the **hub** of the railway system.
- The French railway system has led to closer links with cities outside France.
- The development of settlements is closely connected with the development of the railways and other links.

Settlements and Telecommunications

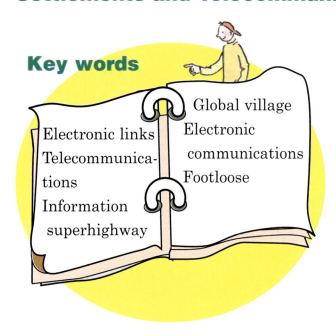

Key words

Electronic links

Telecommunications

Information superhighway

Global village

Electronic communications

Footloose

Road, rail, airways and waterways link people and places. The constant traffic along all these lines of communication helps settlements to grow.

There is yet another way by which places are linked. This is an electronic link using television, fax, telephone, radio, email, computer and video links. **Electronic links** have allowed many settlements to flourish. These **electronic links** are also called **telecommunications**.

The development of the **information superhighway** (the use of the World Wide Web) has changed the way we communicate with each other and the speed at which we do so. This has had an important influence on settlements. It has created the **global village**.

Global village

Electronic links have brought places closer, not in terms of distance, but in terms of time. We can speak now of the **global village**. In terms of communication we are only seconds apart. Japan or Australia are not so far away when a message can be transferred instantly using the telephone or Internet.

A meeting can take place between people living thousands of miles apart using 'tele-conferencing'. The video or telephone link allows people to confer with each other without being in the same place.

Footloose businesses

The growth in **electronic communications** has meant that businesses are more **footloose** than ever. They are freer in terms of where they set up. They can set up away from the main industrial centres and still make the contacts they need for their business.

From Tokyo to Tipperary – not a long way!

This has led to the spread of business over a wider area. Smaller settlements can grow because an Internet or telephone connection is an instant link. Actual distance does not matter.

Distance learning and working

The improvement in information technology (information using the Internet) has allowed people to choose more freely where they want to live and work. People can stay at home and study for their degree using the Internet. This is called distance learning. They do not have to move close to a college or university.

Some people are able to work from home. They can live in a quiet rural settlement, while keeping their link with the city via the telephone and the Internet.

As with the American interstate highway system, people can take jobs or study far from home in terms of distance. This has led to the greater spread of settlement.

Distance learning

Key points

- **Electronic communications** have brought places closer together in terms of time.
- Settlements have grown as their electronic links with the business world have improved.
- **Electronic communications** mean that people can settle at a distance from their work or college.

REVISION EXERCISES

Write the answers in your copybook.

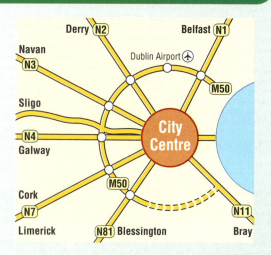

1 Look at the map of the Dublin region.
What purpose does the M50 motorway serve?
- A port tunnel
- A rail connection
- A ring road
- A bus lane

2 Which of the following is not a communication link?
- Roads
- Railways
- Satellites
- Photocopier

3 Which of the following is the lowest class of road in Ireland?
- National secondary
- Regional
- National primary
- Motorway

4 The 'global village' means:
- A village in India
- A worldwide newspaper
- Places in the world are getting closer due to electronic communication
- A community for artists

5 Choose the correct answer for the statements below:
(a) There are over **10,000/58,000** km of roads in Ireland.
(b) In 2002 there was just over **1 million/2.5 million** cars on Irish roads.
(c) The **NRA/RTÉ** is responsible for looking after our roads.

6 Choose the incorrect answer for the statements below:
(a) The main roads in the USA are called **motorways/interstates**.
(b) The TGV is a fast train in **France/the USA**.

7   Describe **two** road projects that have been completed in an effort to improve transport in Ireland.

8   Give **one** way in which an improved road system can help to develop the poorer regions of Ireland. In your answer refer to specifics such as jobs and social life.

9   Describe **three** changes that the interstate system brought to the lives of people in the USA.

10 Give **one** argument in favour of and **one** argument against investing in rail transport over canal transport in France.

11 Explain in detail **two** ways in which an improvement in the transport network in an area has an influence on where people settle. Use named examples in your answer.

12 Electronic communications have attracted modern businesses to small settlements. Give **one** reason for this.

13 Look at the map showing the lines of the TGV.
    (a) Name **three** cities that are linked by this system.
    (b) Explain **one** advantage for cities outside Paris of being linked to the TGV rail system.

14 The chart shows the rise in car ownership in Ireland.
    (a) How many cars were owned in Ireland in 1991?
    (b) Calculate the rise in the number of cars owned between the years 1991 and 2002.
    (c) Describe **one** effect of this increase for car users in Ireland.

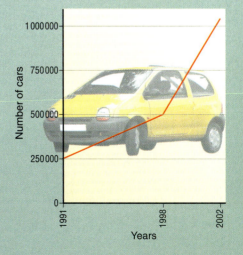

Urbanisation: Changing Patterns in Where We Live

Key words

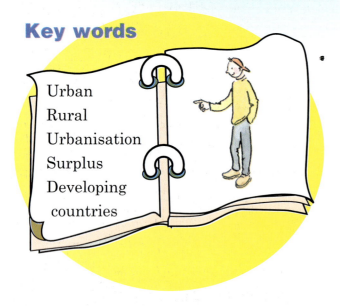

Urban
Rural
Urbanisation
Surplus
Developing
  countries

- There are lots of jobs in the cities and towns.
- There are more houses or flats to rent or buy.
- People are close to services such as schools, hospitals, shops, buses, cinemas, discos and sports centres.

The growth of towns and cities through time

Towns first began as market centres. Later, they became centres of industry. Modern towns have developed into centres of trade, industry, housing and services. Let's look at the development of towns in more detail.

Most people in the world now live in towns and cities. In 1991, for the first time in history, the number of people living in **urban** areas (towns) was greater than those living in **rural** areas (the countryside). We use the word **urbanisation** to describe the increase in the numbers living in urban areas. The increase in urban dwellers has come about because people see many benefits in living and working in towns.

- The first cities in the world developed along the floodplains of rivers. Crops grew well in the rich alluvial soils of these floodplains. This provided a **surplus** of food. Farmers wanted to trade this surplus. Towns developed as market centres.

- Up until the Industrial Revolution most towns were small. During the nineteenth century, with the setting up of factories, towns began to grow very rapidly. Many people moved from rural to urban areas to get jobs in new factories in these large towns and cities.

● As towns grew, they took on a wide range of activities. Towns became centres of administrative functions. These included government offices. Transport and trading activities increased. The towns became centres of social activities.

● By 1850, there were two cities with a population of more than 1 million. These were London and Paris. By 1990, there were 286 'million cities' in the world.

● Since 1950, urbanisation has slowed down in developed countries. In fact, some of the biggest cities in rich countries are losing population as people move away from the city to rural environments.

C	Population of Cities in Millions	
Tokyo	25.0
New York	19.7
São Paulo	16.6
Mexico City	15.1
Los Angeles	15.0
Cairo	14.5
Bombay	12.7
Buenos Aires	12.6
Calcutta	11.1
Seoul	11.0

B	Percentages of People Living in Urban Areas			
	1950	1990	2025 (est.)
Developed countries	53%	74%	84%
Developing countries	17%	34%	57%
World	30%	51%	65%

● Although most people in **developing countries** still live in rural areas, people are moving into towns and cities in very large numbers. Between 1950 and 1990 the urban population in developing countries doubled.

Chart B shows that the percentage of people living in urban areas is growing all the time.

Key points

● Towns developed first as market centres for the produce of the local area.
● Towns grew rapidly with the coming of the Industrial Revolution.
● As towns grew in size they took on a whole range of functions.
● Most of the world's population now lives in **urban** areas.
● Towns in **developing countries** are growing very rapidly.

95

Case Study: The Growth of Dublin

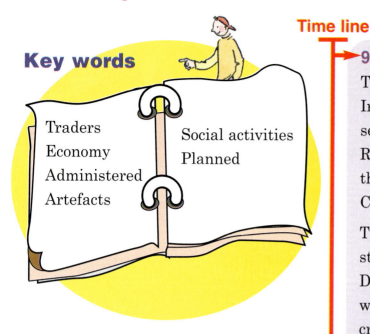

Key words

Traders
Economy
Administered
Artefacts

Social activities
Planned

Let's look at the growth of Dublin. As it grew it developed many functions.

The map shows the main stages in the growth of Dublin. Look at the timeline to see a summary of these stages.

Key		
Viking Dublin	1800–1900	
Medieval Dublin	1900–1970	
Georgian Dublin, 18th century	1970 onwards	

Time line

988 Vikings

The Vikings from Norway settled in Ireland after AD 988. They set up their settlement on a hill overlooking the River Liffey. This settlement was near the present site of Christchurch Cathedral and was also called Dyflyn.

The Vikings were **traders**. There were strong trade links between the town of Dyflyn and Europe. It was a hive of busy workers with a thriving **economy**. Many craftsmen such as shoemakers, carpenters and blacksmiths settled in the town.

1204 Norman (Medieval) Dublin

The Normans arrived in 1204. They overcame the Vikings and built Dublin Castle on the site of the old Viking stronghold. From here they **administered** their control over the town. They built a stone wall with look-out towers around the city. The town grew inside these walls. There was terrible overcrowding within the walled city.

The mayor and council administered all aspects of the town's life. Most of the civil servants were clergymen. Traders imported goods from far and wide, as **artefacts** found at Wood Quay show.

1600 Expansion

During the seventeenth century the city grew. The Grand and Royal canals were built and these linked the city to the Midlands. The canals brought greater trade to Dublin as goods passed through its port. More industries were set up in the city.

1700 Georgian Dublin

During the eighteenth century beautiful houses and squares e.g. Mountjoy, Fitzwilliam and Merrion Squares were built for the rich of the city.

Social activities became more varied and this brought more people into the city. To cope with the increasing growth of the city streets were carefully **planned**.

1800 Growth of the Suburbs

During the nineteenth century many of the Georgian houses became tenements. Previous owners returned to London. Often over 100 people lived in one house. They had no inside toilets.

Dublin was growing because there were jobs in the city and a good social life. People left country areas for the city.

During the Famine more people migrated to Dublin. Railway lines and roads were built and these allowed the city to spread. Wealthy people went to live in the new suburbs. Villages like Rathgar became part of the city.

1900 Planned Housing Schemes

In the early part of the twentieth century, the city of Dublin faced a crisis in housing. Most of the jobs in government, factories and transport were based in Dublin as were social activities. This caused the city to grow even more. New housing schemes had to be planned such as those at Drumcondra. Public transport now served these areas.

1970 New Towns

In the later part of the twentieth century, new towns such as Tallaght, Blanchardstown and Lucan were built on the western side of the city. Land that had once been farmland was now developed for housing. Four local authorities were set up to **administer** the different areas of the city.

Key points

- Dublin has grown from a small walled town in the tenth century to a city of over 1 million people.
- Trading is one of the main functions of the city.
- As the city developed it took on important **economic**, **administration** and **social** functions.

REVISION EXERCISES

Write the answers in your copybook.

1 Urbanisation is:
   - The setting up of factories on the edge of cities
   - The growth of towns and cities
   - The movement of people to the countryside
   - The spread of shopping centres

2 Choose the correct answer from the following statements:
   (a) The first cities grew along the floodplain of
       rivers/mountainsides.
   (b) Towns began to spread rapidly after the
       technological/Industrial Revolution.
   (c) Urban population **trebled/doubled** between the years 1950–90.

3 Match each letter in column X with the number of its pair in column Y.

X	Y	ANSWER
A Urbanisation	1 Market centres	A =
B Rural areas	2 Dublin	B =
C First towns	3 Growth of towns and cities	C =
D Vikings	4 Growth of the countryside	D =

4 Match each letter in column X with the number of its pair in column Y.

X	Y	ANSWER
A Early Viking site	1 Merrion Square	A =
B Built by the Normans	2 Near Christchurch Cathedral	B =
C Dates from the eighteenth century	3 Tallaght	C =
D New town	4 Dublin Castle	D =

5 Explain **two** causes for the growth of cities in the developing world.

6 Name **two** main differences of activities between urban and rural areas.

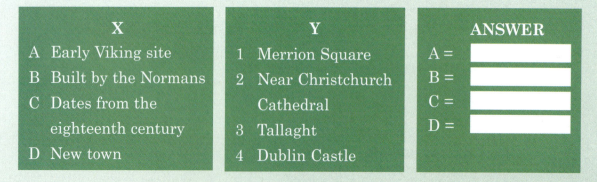

7   Describe **two** benefits of living in urban areas.

8   Name the county council responsible for your area. Briefly describe **two** services that it carries out for the area.

9   Draw a map of the Dublin region. Show, using different colours, how it has expanded from its earliest days to the present day. In what direction is most of the new development taking place? Explain **one** reason why this is so.

10   Look at the table showing the population of cities in millions.
  (a) Using the information in the table, draw a bar chart showing the population in the cities.
  (b) Using the same table, list the cities that are located in the developing world.

Population of Cities in Millions	
Tokyo	25.0
New York	19.7
São Paulo	16.6
Mexico City	15.1
Los Angeles	15.0
Cairo	14.5
Bombay	12.7
Buenos Aires	12.6
Calcutta	11.1
Seoul	11.0

11   The following information looks at the percentage of people living in towns or cities throughout the developed and the developing world. It uses information from the years 1950 and 1990 to suggest the likely percentages of world population that will live in urban areas in the year 2025.

Levels of Urbanisation in 1950, 1990 and 2025			
	1950	1990	2025 (est.)
Developed countries	53%	74%	84%
Developing countries	17%	34%	57%

  (a) Draw three bar charts showing the information for each of the years.
  (b) Describe the main pattern that you see when you examine the three charts.

12 Functional Zones in a City

Key words

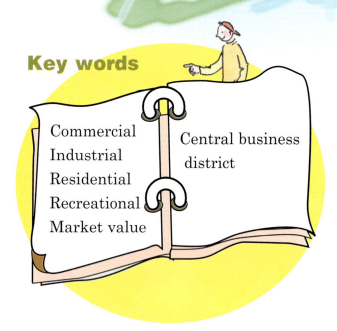

Commercial
Industrial
Residential
Recreational
Market value

Central business district

All sorts of activities, also called functions, go on in a city. In the city you will find:

- Shops and offices (**commercial** functions)
- Factories (**industrial** functions)
- Houses and apartments (**residential** functions)
- Playgrounds, theatres and cinemas (**recreational** functions).

These functions can often be found in the same part of the city. Where this happens, we describe it as a functional zone. We can refer to a factory zone, a shopping zone and a housing zone.

The part of the city where you find these functional zones depends on a number of factors. These include:

- Age of buildings
- **Market value** or cost of land
- Communication links (water, rail or road).

Let's look briefly at each of these factors.

Age of buildings

The oldest buildings are found at the centre of the town. This is where the town started and the place from which it grew.

Older buildings in the centre of a European city

Market value of the land

The most valuable land is at the centre of the town. The land is cheaper as you move towards the edge. This has to do with the demand for land in the market. More people want to be near the centre of the town. This pushes the price up. The value or cost of the land has an influence on how that land is used.

Sold to the highest bidder!

Communication links

Land costs more if it is within easy reach of routeways. Businesses can afford the high rents. They set up near main routes.

Zones in a city

Because of the influence of factors such as the age of buildings, the cost of land, and communication links, there is usually a pattern in where you find certain activities in a city. The model shown in the diagram gives you an impression of this pattern.

Let's look in a little detail at the zone at the centre of the town or city.

They must make the best use of this space. That is why you find the tallest buildings at the centre of the town. Building up is cheaper than building out. People live in apartments over shops or offices in the central business district.

During the day the CBD is very crowded. This is due to the high number of shoppers, and shop and office workers. It becomes less busy at night as most workers and shoppers return home to the suburbs.

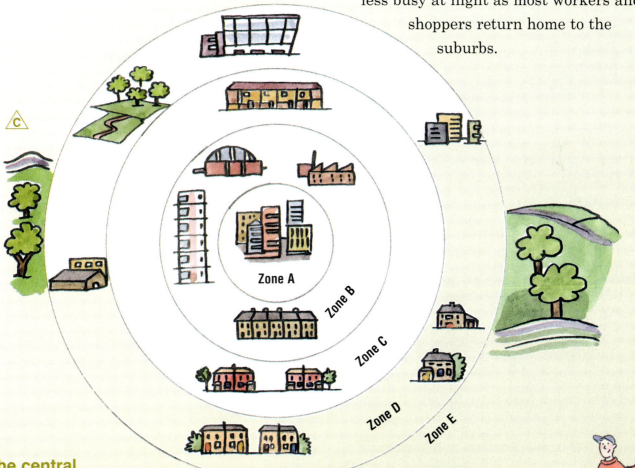

Zone A

Zone B

Zone C

Zone D

Zone E

The central business district (CBD)

The **central business district** (CBD) (zone A on the diagram) is at the heart of the town. It is the main office, restaurant, theatre and shopping area. The value of the land is highest here.

The CBD is the most accessible place. The main roads lead to the centre of the town. It is the best place for business. For this reason, people pay high rents to set up here.

Key points

- There are zones of different land uses in towns. These can be shown on a model.
- Factors such as the age of buildings, the value of land and closeness to routeways have an influence on land use.
- The **central business district** is found at the centre of the town.

Zones in the City – A Closer Look

Key words

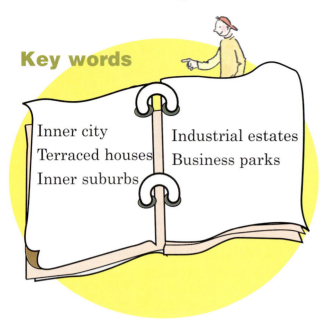

Inner city
Terraced houses
Inner suburbs

Industrial estates
Business parks

Zone B inner-city housing

Zone B – changing land use

As you travel beyond the edge of the central business district you move into a zone of changing land use (see zone B in diagram C on p. 101). It is a zone where redevelopment is taking place. Apartment blocks or modern town houses are replacing old factories.

In the early nineteenth century, when towns were small, factories making clothes and processing food were found close to what is now the **inner city**. They were built next to rivers and canals, railways and roads.

Between the 1950s and the 1980s these industries moved from the inner city to edge-of-city sites where there was more space to expand. Land is cheaper there.

Other firms that need less space, such as bakeries, dairies and printing works are still found close to the centre.

Zone C – housing zone

Zone C is on the edge of the **inner city**. In this zone you would find a mixture of old, cramped housing and newer dwellings.

Over 100 years ago, long, straight rows of **terraced houses** were built close to the nearby factories. People had to be near their place of work because transport links were not good at that time.

By the 1950s, many of these inner-city houses had become slums. Large areas were flattened by bulldozers and replaced with new high-rise blocks of flats.

Sometimes, older housing was improved, rather than replaced. Bathrooms, kitchens, hot water and indoor toilets were added.

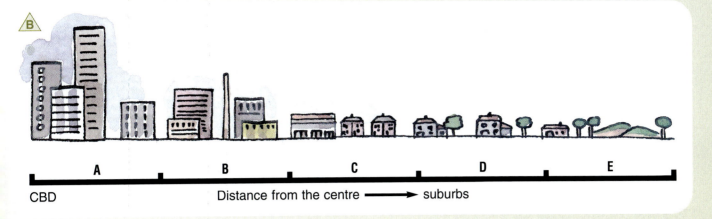

B

A   B   C   D   E

CBD   Distance from the centre ——▶ suburbs

Zone D – housing zone in the old inner suburbs

As you continue to move out towards the edge of the city, you move into the old **inner suburbs** of the city. The houses here were built in the 1920s and 1930s.

Houses in zone D are not as cramped as houses nearer to the city centre. Land is cheaper here and the houses are bigger. They have front and back gardens and three or four bedrooms. These houses were once at the edge of the city, but the city has grown out beyond them. They are usually served by small local shopping areas.

Zone E – modern suburbs

Zone E is found at the edge of the city. These are the modern suburbs. You will find a mixture of housing estates, **industrial estates**, **business parks** and large shopping centres in this zone. This zone includes the many new towns that have grown around the edges of cities. These are areas of planned housing and services.

Industrial estates and modern business parks are located within zone E. They have been built on large areas of quite cheap land close to main routeways.

Large shopping centres or malls have been built at the edges of towns because people want to avoid the traffic congestion found on roads heading into the city centre. Parking is good in the shopping centre. There are a wide variety of shops in the shopping centre. The fact that you can find these shops under the one roof is also attractive to shoppers.

Parks and open spaces in the city

The planners in most cities leave some open spaces for leisure activities. Mindful of the pressure to use this land for housing or industry, the city planners often protect this land by law. It is important to set aside some green spaces in the middle of the urban jungle.

Key points

- Outside the CBD there is a zone of changing land use.
- The industrial zone is now at the edge of the city.

The Functional Zones of London

Key words

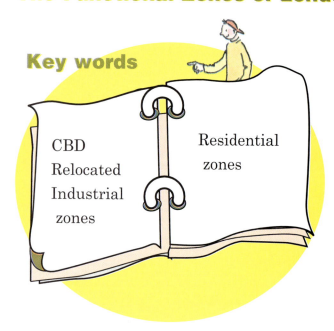

CBD
Relocated
Industrial
zones

Residential
zones

London is the capital city of England and seat of the government of Great Britain and Northern Ireland. It is in south-east England and lies on the banks of the River Thames. It has a population of over 7.2 million. It is a vast city stretching some 50 km from one side to the other. It is the major port and trading area of Great Britain.

Central business district

The West End is at the centre of London. It is a major shopping area where many of the department store chains have branches. This is where you would find Covent Garden, Trafalgar Square and Oxford Street.

'The City' is the business zone within the **CBD**. Modern office blocks of glass and steel stand side by side with old buildings. The Bank of England, the Royal Exchange, Lloyds and headquarters of other major businesses are in 'The City'.

The Docklands is a redeveloped area close to the city centre. Many businesses have **relocated** here from central London. These include the newspaper industry that has moved from Fleet Street. Canary Wharf, a newly developed area in this zone, has modern office and apartment blocks.

There are areas in London that you would link with particular activities. Let's look at functional zones within the city of London.

A

London

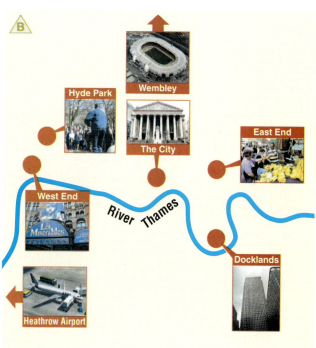

B

Hyde Park

Wembley

The City

East End

West End

River Thames

Docklands

Heathrow Airport

London – functional zones

Industrial zones

The factory zone is found, for the most part, towards the edge of the city. These areas include Park Royal, Wembley and the area around Heathrow Airport. Land is cheaper here and there is access to air transport.

Residential zones

The East End is an inner suburb of London. This area is home to many immigrants including Irish, Jews, Pakistanis and Indians. There is a mixture of working-class houses, small businesses, shops and markets. Many of the houses are terraced. Think of *EastEnders* on TV!

Good public transport, such as the underground, has allowed people to live in bigger houses in the outer suburbs. Wimbledon, famous for its tennis and its football team, is one of the newer residential areas of London.

Open space and parks

About 40% of London's total area is made up of green spaces and parks. Many of these are well known such as Hyde Park, Regent's Park and St James's Park.

London and the model of urban land use

As in the model of urban land use, there are zones of different land use in London. Business is at the centre and the factory zone, for the most part, is at the edge. As in the model, houses are found in all zones. The oldest and smallest houses are near the centre. The larger houses are at the edge. Parks are found throughout the city.

Regent's Park

Oxford Street

Canary Wharf Tower

Key points

- There are zones of different land use in London.

Intensity of Land Use in a City

Key words

Land value
Intensive
Floor space

The cost of land is dearest in the centre of a city. The cheapest land is found at the city's edge. This is called a difference in **land value**.

High land values at the city centre

Many businesses want to be close to the centre of the city. This is the busiest place, with access to the most customers. The most money can be made here. Businesses compete with each other for this space. It is like an auction. Competition between buyers pushes the price up. This explains why the cost of land increases towards the city centre.

SOLD!

Land value and land use

If you pay a high price for land you will want to make the most of that space. Building up rather than out is an **intensive** use of space. This is why you find the highest buildings at the centre of the city centre.

Let's look in a little more detail at how buildings are used in the city centre.

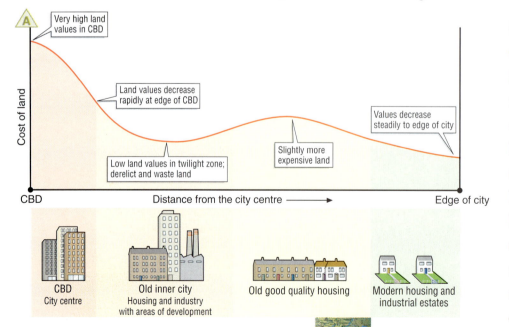

Land use at the centre of the city

Shops, bars and restaurants are found on the ground floors of buildings in the city centre. These kinds of businesses have a high turnover of customers. They make a lot of money and can afford the high rents on this land.

The higher floors in buildings close to the city centre are used as office spaces. Rents are high here.

Housing

There is some housing in the city centre. Houses tend to be small and many do not have front or back gardens. This has to do with the high cost of land. A bigger house in the city centre would take up more space and would cost a lot.

You find multi-storey apartment blocks nearer to the city centre. Building high-rise dwellings makes good sense in this case. If you build apartments you are getting the most out of this valuable land.

In general, you find more office and retail land use than housing land use in the city centre. Businesses are better able to afford the high cost of the land.

City-centre housing in Cork

Land use at the edge of the city

As you move out towards the edge of the city, the value of the land falls. This has to do with the lower demand for land at the edge of the city. Lower demand pushes the price down.

Factories

Factories need a lot of **floor space**. Often they use big, heavy machines. It is not suitable to have these machines on different floors. Some factories use plenty of raw materials that have to be stored nearby. They also need large areas where they can park while loading and unloading their goods. You find factories at the city's edge where the cost of land is low.

Shopping centres

Shopping centres at the city's edge take up more space than shopping areas near the city centre. They can do this because the land is cheaper at the edge. You find large, open-air car parks, rather than multi-storey car parks.

Housing

Houses at the edge of the city tend to be bigger than in the city centre. They will probably have a front and a back garden. This is possible at the city's edge where land is cheaper than in the centre.

Big houses with plenty of green areas at the edge of the city

Key points

- Land is used **intensively** near the centre of the town rather than at the edge.
- Land use is influenced by the **value** of the **land**.

Case Study: Intensive Land Use in Dublin's Docklands

Key words

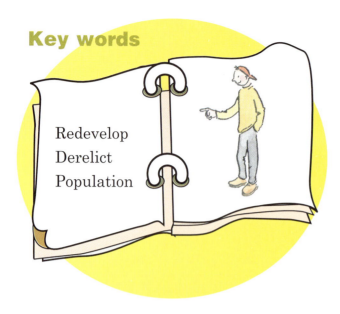

Redevelop
Derelict
Population

The city planners decided to **redevelop** an area of urban decline close to Dublin's city centre. This area, the Docklands, had fallen into a state of disrepair when port activities moved further down the river closer to the sea. Many people are now familiar with this area from going to concerts at the Point. The idea was that people would settle here and jobs and social activities would be found side by side.

The Docklands' development

The Docklands' development is close to the heart of Dublin city. There are some 520 hectares (1,300 acres) of prime river-side land here. Old **derelict** sites were cleared and a programme of rebuilding was planned. New offices, shops, hotels, homes and leisure facilities now stand on this once vacant site. Here you will find the International Financial Services Centre (IFSC). There are over 370 international financial companies here employing nearly 14,000 people.

Unlike many American cities, skyscrapers are not a feature of the Dublin skyline.

While it is agreed that Dublin should not become a city of skyscrapers, it is also agreed that it should not become a sprawling city. Higher densities are therefore allowed in some places so that the city does not spread out too far into the countryside.

New buildings such as the towers of light on George's Quay, directly opposite the Custom House, show intensive land use

Redevelopment in the Dockland area

Redevelopment zones in the Dublin Port area. Map courtesy of Dublin Docklands Development Authority

By the year 2012, the **population** of the Docklands will have increased by 25,000 to 42,500. In time, over 11,000 new homes and up to 40,000 new jobs will be provided in this area.

Other projects are planned for the Docklands on both sides of the River Liffey.

The map shows how close the Docklands' area is to the centre of Dublin. The demand from businesses to set up in the Docklands area has pushed up the price of the land. This in turn has meant it has had to be used intensively.

Leisure time in redeveloped areas

Key points

- The Dublin Docklands is an area which has been **redeveloped**.
- Land is used intensively because it has a high value.

City Housing

Key words

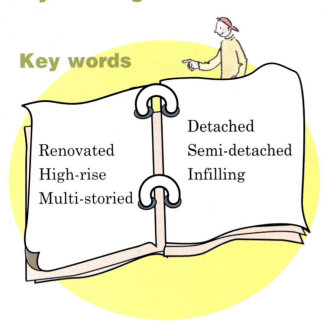

Renovated
High-rise
Multi-storied

Detached
Semi-detached
Infilling

Let's look at how the type, age and quality of housing can vary within a city.

Case Study: Dublin
Age and type of housing in the inner city

Some of the oldest houses in the city are in the city centre. Often these are long, straight rows of terraced, two-storied houses. They are small houses, sometimes with two rooms upstairs and one downstairs. Many were built 'back to back'. They have a small backyard but no front garden. They may not have an indoor toilet or bathroom. Some of these houses are more than 100 years old.

This is high-density housing i.e. many houses are crowded into a small space. The houses were built close to the old factories where the workers were within walking distance of their jobs. The houses competed with the factories for the space and the factories won. They took up the biggest space and so the houses were cramped into a smaller area.

Inner-city housing

With the help of government grants and tax breaks, some of the older houses are being bought and **renovated** by well-paid office workers. They want to live close to their places of work.

In some parts of the inner city, large areas of old houses, warehouses and factories have been flattened by bulldozers. Town houses have been built in their place. You also find modern, **high-rise (multi-storied)** apartment blocks close to the city centre.

Age and type of housing in the suburbs

As you move outwards from the city centre the age and type of housing generally changes. Land becomes cheaper as you move out from the city centre. For this reason, houses are larger than in the inner city. They have three or four bedrooms. They have gardens at the front and back. Some are **detached**, **semi-detached** or are in short terraces of four or six houses.

As you move out from the city towards the outer suburbs the houses generally become less old. These estates consist of a mixture of public (local authority) and privately built houses.

There is a lot of **infilling** in suburban areas. Any open space is targeted. People are selling parts of their gardens for new houses, particularly on corner sites. Multi-storied apartment blocks are springing up in areas that were once only used for two-storied family houses. This is a major change in the landscape of cities. New modern dwellings are mixed in with older dwellings.

Infilling

New towns

New towns have been planned and built on the outskirts of cities such as Dublin. New towns, such as Tallaght and Blanchardstown, have a mixture of house types. You find the old terraced houses that were once part of the original village settlement, new detached and semi-detached houses and multi-storied apartment complexes.

Old terraced housing in Tallaght

Quality of housing in cities

The quality of housing is not the same throughout the city. In wealthy areas the houses and gardens are large and the streets are lined with trees. This type of housing gives a lot of privacy to the residents.

In less well-off areas the houses seem to be on top of each other. There may be endless lines of houses all looking alike and few trees. There is little privacy.

New housing developments in the city show a vast improvement in quality. The areas are landscaped with mature shrubs and trees.

Mixed housing complexes

Key points

- The quality, type and age of houses varies within the city.
- There is a pattern of mixed housing in the city centre and suburbs. There are modern, **multi-storied** apartment blocks springing up alongside older, two-storied dwellings.

REVISION EXERCISES

Write the answers in your copybook.

1   In cities the main area or zone for business and shopping is called:
   - The suburbs
   - The central business district
   - The business park
   - The industrial area

2   The aim in the illustration can best be achieved by:
   - Building more factories
   - Opening more shopping centres
   - Better landscape planning
   - Increasing car ownership

3   The statement which best describes the situation in cities in developed countries is:
   - In the suburbs land values increase, so buildings are higher
   - In the city centre land values increase, so buildings are further apart
   - In the suburbs land values increase, so buildings are further apart
   - In the city centre land values increase, so buildings are higher

4   The price that is paid for a site is called:
   - Land value
   - Site value

5   Match each letter in column X with the number of its pair in column Y.

X	Y	ANSWER
A Main area for shops and offices	1 Recreation	A =
B New factories	2 Commuters	B =
C New residential areas	3 CBD	C =
D Football pitch	4 Suburbs	D =
E People who travel to work	5 Industrial estate	E =

6 Look at this drawing showing how the average height of buildings changes between a city centre and the suburbs. If you were moving in the direction of the arrow XY, would you be:

- Moving towards the city centre?
- Moving away from the city centre?

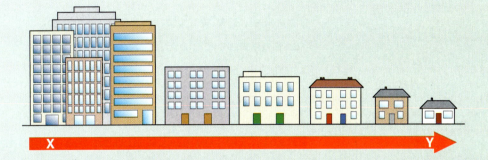

7 Clearly describe **two** ways that you would recognise the CBD of any city. You may answer with a clearly labelled diagram.

8 Explain, using examples from your own town, how **two** of the following factors influence how and for what land is used.

- Age of buildings
- Market or present value of the buildings
- Nearness to good communication links

9 Explain **one** way houses further away from the city centre are different to those houses found in inner-city areas.

10 Describe **two** different types of land use zones found in a named city that you have studied. You can take London or your own town as your named city.

11 Give **one** reason why factories need lots of ground floor space.

12 Draw a diagram showing the main differences between houses in the city centre and those further out. Refer to height and width of houses.

13 In the case of **one** city that you have studied describe the differences in the way land is used from the city centre to the edge of this city.

13 Patterns of Daily Movements in Cities

Key words

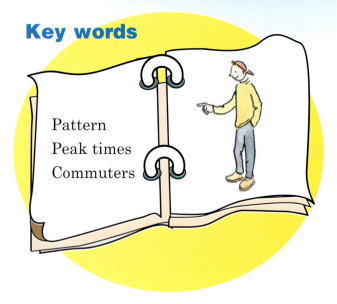

Pattern
Peak times
Commuters

The movement of people to work or school in our cities makes a pattern. A **pattern** is something that is repeated from one day to the next.

● There is a pattern in terms of the time of movement.
● There is a pattern in terms of the place and direction of movement.

In Ireland, people generally set off for work or school sometime between 7am and 9am. The country is on the move at this time.

Since businesses are usually found close to the CBD and housing is found further out, traffic flows along the main routes will be greater in the direction of the city in the early morning.

Between 4pm and 7pm, the country is on the move again. People are going home. There is a huge flow of traffic out of the city along the main routes.

The times when there is more traffic moving along the road are called **peak times**.

Think about the movement in your own locality. Do you notice that in the morning it is busier in one direction and in the evening it is busier in the opposite direction?

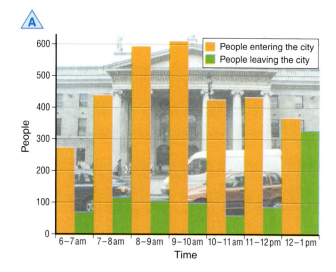

A

People entering the city
People leaving the city

Pattern of movement in towns

The people responsible for transport in the city need to know about the pattern of movement there. They need to know which routes are busiest and the times they are busy. This allows them, for example, to plan to have more buses running during peak times, when people travel to and from work or schools. They will not need to put on so many buses when the morning and evening rush is over.

Commuters, the travelling public, also need to know the pattern (the times and places) of movement. These will affect the time they need to make the journey.

Increase in car ownership

The 2002 Census showed that cars are the principal means of travel for Irish workers. More people drove to work by car in 2002 than ever before. Over 1 million households had at least one car in 2002. This showed an increase of 330,000 from 1991.

People also travelled further to work, many facing journeys of over 80 km. Over 66% of workers used private cars, whereas the number using public transport is now only 8.8%.

But it is not only workers who travel each day. Take a look at the figures for primary school students. These show the percentage (%) who walked and those who got lifts.

D Primary School Students		
	Car	Walk
1981	19.7%	47.3%
2002	50%	26%

Nowadays, more and more people are travelling long distances to work or school. Most people travel to work by car and will set off roughly at the same time. It is no wonder that the problem of traffic congestion is getting worse!

Commuters

The population in the 'commuter belt' around cities such as Galway, Limerick, Cork and Waterford is growing at an alarming rate.

However, as we have mentioned before, the biggest increase in this 'commuter traffic' is happening in areas outside the Greater Dublin Area. People living in these areas commute daily into Dublin for work.

These commuters are early risers. Many are up by 6am to face the long journey to work. Irish roads have improved greatly over the last 20 years, but the large numbers of cars clogging up the roads is cancelling out the benefits of this improvement.

'There is a tail back of 4 kilometres along the Naas by-pass'. AA Roadwatch

Key points

- There are **patterns** in the way people move about in cities.
- **Patterns** of movement in time and space are repeated from day to day.

Managing Traffic in Our Cities

Key words

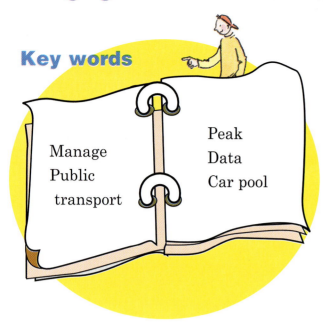

Manage
Public
transport

Peak
Data
Car pool

The Luas is a light rail system. Look at the map of the Luas lines.

You will notice that the Red Line and Green Line do not link up. There is a 15-minute walk between Abbey Street on the Red Line and St Stephen's Green on the Green Line. However, tickets bought for a journey starting on one line and finishing on the other are valid for the whole journey. This makes it more appealing to passengers. It is likely that further lines

Because the pattern of movement of people in the city is known, it is possible to come up with a plan to **manage** this movement. In this way, the problems of traffic congestion can be lessened.

ZONES

A

Museum · Smithfield · The Four Courts · Jervis · Abbey Street · Busáras · Connolly

Heuston
James's
Fatima
Rialto
Suir Road
Goldenbridge
Drimnagh
Blackhorse
Bluebell
Kylemore
Red Cow
Kingswood
Belgard
Cookstown
Hospital
Tallaght

St Stephen's Green
Harcourt
Charlemont
Ranelagh
Beechwood
Cowper
Milltown
Windy Arbour
Dundrum
Balally
Kilmacud
Stillorgan
Sandyford

Red 4 · Red 3 · Red 2 · Central 1 · Green 2 · Green 3

Improving public transport

One solution is to get people to use **public transport** rather than private cars. For this to happen, we must improve rail and bus links. Look at the following steps which have been taken.

will be added to the Luas network providing better links between places in the city.

Rail and light rail links

Rail links between Dublin and the commuter belt have been improved. More people are now commuting by train e.g. Arrow and DART lines.

A network of light rail links has been built from the outlying areas of Tallaght and Sandyford into the city. This is the Luas line. *Luas* is the Irish word for *speed*.

The Luas is Dublin's new tram

Quality bus corridors (bus lanes)

Traffic management

Bus lanes are for the sole use of buses and taxis. Journey times are speeded up. At **peak** hours, **public transport** takes you to the city centre in a faster time than it would if you travelled by car. It may not be as stressful either. It is hoped that in the future there will be fewer private cars on the roads.

These improvements in **public transport** will lessen the problems of traffic congestion at peak times.

Controlling traffic flows

You have learned that there are times during the day when traffic jams are worse than others. A driver can be sitting at traffic signals for several minutes. There is a way of reducing this problem.

The SCATS (Sydney Co-ordinated Adaptive Traffic System) is used in Dublin City. SCATS gathers **data** on traffic flows where roads meet. This data is fed through the traffic control signal box to a central computer. The computer can then control the traffic light timings as the traffic flow changes. If traffic is heavier in one direction the traffic lights will stay green for longer there. The SCATS system is also used in many other cities including Hong Kong, Sydney and Melbourne.

Websites advertising car pools

The large number of cars on our roads at peak time is making the problem of traffic congestion worse. Having a **car pool** can cut the numbers of cars on the roads. This would be really useful at the busy times when everybody seems to be moving in the same direction.

In a car pool people who travel the same route each day agree to travel together and share the cost. If two or more people were travelling in one car, this car could be allowed to use the bus lane. Cities such as San Francisco have successfully introduced this system. Lately, a number of websites have advertised this service.

Key points

- During **peak** times of traffic the flow can be speeded up by: using **public transport**, monitoring traffic flows, controlling traffic light changes and car pooling.

REVISION EXERCISES

Write the answers in your copybook.

1   Examine the bar chart. Which of the following statements best describes the information shown here?

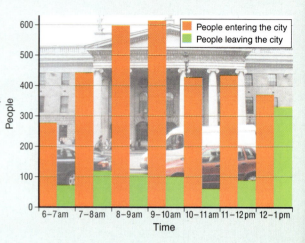

- Most traffic enters the city between 6–8pm
- Most traffic leaving the city does so between 9–11am
- The busiest time for incoming traffic is between 8–10am in the morning
- The quietest time for outgoing traffic is between 12–1pm

2   The main means of transport for Irish workers is:
- Cars
- Bicycles
- Trains
- Buses

3   Match each letter in column X with the number of its pair in column Y.

X	Y	ANSWER
A SCATS	1 Luas	A =
B Quality bus corridor	2 Traffic Control System	B =
C Light rail	3 Daily movement of people	C =
D Commuters	4 Bus lane	D =

4   Give **two** reasons why traffic congestion happens so often in our towns and cities.

5   Briefly describe **two** methods which have been used to ease such congestion.

6   Imagine that you work in the traffic planning section of your local county council. Name **two** solutions you would use to solve the traffic problems in your area. Clearly explain why you chose those solutions.

7 Describe **one** advantage of the Luas line over the DART line for easing traffic congestion.

8 Many people support the greater use of bicycle lanes in our cities. Suggest **one** reason why so few of these lanes are found in Irish cities.

9 One solution offered to ease traffic congestion in our larger cities is to build car parks on the edge of towns and cities. Suggest **one** argument in favour and **one** argument against this solution.

10 Examine the map below.
   (a) Name **three** counties where a high percentage of workers travel 32 km or more to work.
   (b) Give **one** reason why this pattern exists in these counties.

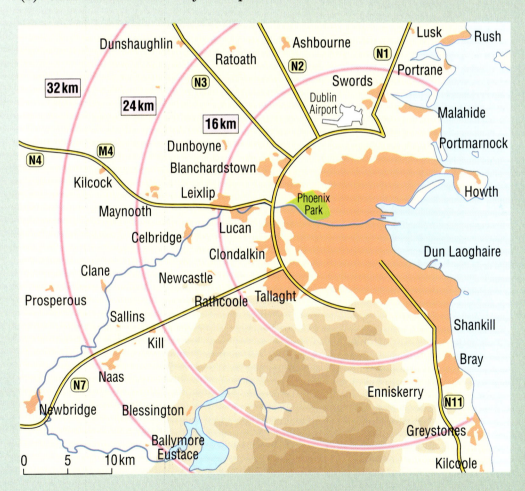

Key words

Environments
Urban decline
Urban sprawl
Crime rate
Pastureland

A number of problems have arisen in modern city **environments**. These problems include:

- Crime
- **Urban decline**
- **Urban sprawl**.

Crime

Each year, the Garda Síochána publish a report on crime. The report gives a breakdown of the crimes committed in the country. It describes the types of crimes that are committed. It also compares crime rates in different regions.

Types of crime

One type of crime is called non-headline crime. These are offences such as traffic offences and minor acts of vandalism. These types of crimes are not usually reported in the newspapers.

Other, more serious crimes, are called headline crimes. Headline crimes are

crimes like murder and violent assault. These are the crimes you hear about in the media.

Rates of crime in different regions

To compare the level of crime in one place with that of another place you need to know the **crime rate**. When we know the total population in each region, and the number of crimes committed in each region, we can work out the rate of crime per head of population. We can then compare crime rates in the different regions.

According to *An Garda Síochána Annual Report 2002*, crime rates are higher in cities than in rural areas.

The bar chart B shows us that the most urbanised area, the Dublin Metropolitan Area (DMA), reported the highest rate of headline crimes. It shows that for every 1,000 people who live in the DMA, an average of 47 headline crimes were committed. This is more than twice the figure for the more rural western region.

The figures show that in the highly-populated, urban areas of Ireland crime rates are higher than in rural areas.

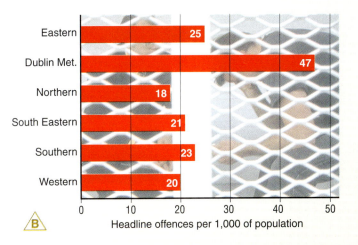

Eastern — 25
Dublin Met. — 47
Northern — 18
South Eastern — 21
Southern — 23
Western — 20

0 10 20 30 40 50

B Headline offences per 1,000 of population

• The quiet life of country people is spoiled.

• As the city spreads, people often have to travel long distances to get to work. This means a huge increase in the number of cars on the road. This causes pollution: air, noise and visual.

In Ireland we have a lot of urban sprawl around our cities. Dublin has now spread out as far as places such as Dunshaughlin in Co. Meath and Kilcock in Co. Kildare. These towns are now commuter zones for Dublin.

Zones of decline

Another problem in modern cities is the problem of **urban decline**. There are parts of our cities that are still very run-down. Buildings are in a derelict state with doors and windows boarded up.

• Often squatters have moved into these zones of decline bringing a whole set of problems with them such as homelessness and drug abuse.

• Many locals move out when their area becomes run-down. Some older people may not have money to move away. Life can be difficult for them as the old community is broken up.

• These zones of decline can become no-go areas in the city. The low population density makes them unsafe to walk about in. A zone of decline often has high levels of crime.

Commuter zones around Dublin

Urban sprawl

Urban sprawl is the spread of towns and cities into the surrounding countryside. This is a problem in modern cities for a number of reasons.

• Valuable farmland is swallowed up when houses and roads eat into the countryside. For example, this brought a 4% decrease in **pastureland** in Ireland during the 1990s.

Key points

• **Crime rates** are higher in urban areas than in the countryside.

• Zones of decline are run-down areas that are unattractive and sometimes dangerous.

• **Urban sprawl** causes problems for people living in the city and the country.

REVISION EXERCISES

Write the answers in your copybook.

1 Urban sprawl is:
- The spread of a city into the surrounding countryside
- A rapid increase in the number of tall buildings in a city
- The growth of a city's traffic
- The fast growth of a city's population

2 The statement which best describes the situation in the cartoon:
- More land is needed for farming
- Urban development is carefully planned
- Urban development rolls on at the expense of farmland
- Suburban housing estates are located at the edge of cities

3 Match each letter in column X with the number of its pair in column Y.

X	Y	ANSWER
A Non-headline crimes	1 Areas of derelict buildings	A =
B Headline crime	2 Murder	B =
C Urban decline	3 Farmland is used for housing	C =
D Urban sprawl	4 Traffic offences	D =

4 Many of the present problems in our towns and cities are said to be due to the speed at which changes have taken place. Large housing estates have been built on the edge of cities and factories have moved nearby. This creates problems in the older areas of the inner city. Describe **two** of these problems.

5 Suggest **one** solution to the problem of crime in inner-city areas.

6 Imagine that you lived 50 km from Dublin. Describe what changes you have noticed as builders buy land to use for housing.

7 Suggest **one** solution to the problem of urban sprawl.

8 Imagine that you are a parent of a young family. Give **one** argument in favour of and **one** argument against living with your family in the inner city.

9 Look carefully at the cartoon opposite.
(a) What is the message of this cartoon?
(b) Suggest **one** solution to the problem of urban sprawl.

10 Look at the bar chart showing headline crimes for 2002.
(a) Name **two** headline crimes.
(b) Calculate the difference in the number of crimes committed between the area with the highest and lowest number of crimes.
(c) Give one reason why the Dublin Metropolitan area reported the highest rate of headline crimes.
(d) When looking at figures for different areas why is it more useful to calculate the crime rate rather than the actual number of crimes?

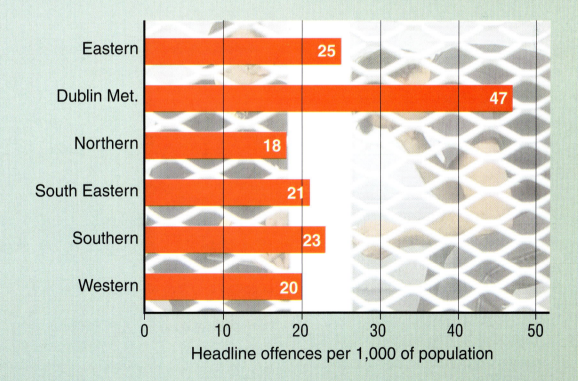

Headline offences per 1,000 of population

15 Urban Renewal and Redevelopment

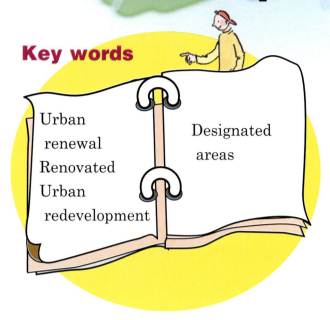

Key words

Urban renewal

Renovated

Urban redevelopment

Designated areas

What is urban renewal and redevelopment?

Over time, parts of a city can become very run-down. When **urban renewal** takes place, old and run-down buildings are **renovated** or replaced with newer buildings. These buildings are then used for housing.

Besides the renewal of housing, there are other types of **urban redevelopment** in the city. New shops and offices are built on cleared sites.

The government wanted to breathe new life into inner-city areas that were run-down. This had happened in cities such as Dublin, Waterford and Cork. The government wanted to rebuild communities here. To do this, they encouraged people to set up businesses and to come and live in the renewed areas. Many urban renewal projects were undertaken during the 1980s.

Urban renewal plans

Town planners came up with plans for inner-city areas.

- There would be small businesses in these areas so that there would be jobs for local people.
- There would be homes nearby, such as townhouses or apartments.
- Grants or tax breaks were offered to developers to invest in housing and business projects in parts of the city. These are called **designated areas**.

Inner-city renewal projects have been very successful. Let's look at one success story – an urban renewal scheme in Dublin.

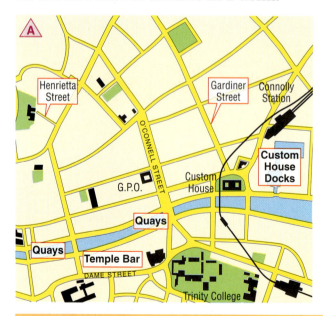

Location and Size of Designated Areas in Dublin City, 1986

Designated Areas	Acres
Gardiner Street	91
Quays	68
Henrietta Street	2.5
Custom House Docks	27
Temple Bar	28

Case Study: Temple Bar District of Dublin

The eastern end of Temple Bar was part of the medieval city of Dublin. The area to the west of Parliament Street was mainly built up in the eighteenth century. It was an important centre for housing, trade and crafts.

Look at map A. It shows how Temple Bar lies at the heart of Dublin. In the twentieth century much of Temple Bar started to become run-down. Many of the shops and small businesses closed. People moved out of the area to new housing estates in the suburbs. However, some small clothing businesses remained.

CIE, the transport company, bought some of the derelict buildings. They planned to build a central bus station in the area. In the meantime, rather than leave the buildings empty, they leased them at low rents to small restaurants, art galleries, second-hand clothes shops and others. Many young people, including artists, came to live or work in the Temple Bar area.

The area began to take on an identity as a cultural quarter.

Under pressure from local people the government stopped CIE's plan for the bus station. Instead, the government decided to redevelop Temple Bar, using a company called Temple Bar Properties, which was set up in 1991 to manage the project.

Ten years later the area has been completely transformed. There are many apartments, shops, restaurants, pubs and businesses. There are art galleries, theatres and cinemas. Many artists live and work in the area. Its medieval history has been respected. The Viking site at the eastern end of the area was excavated and a Viking centre was built to display this history. Overall the project to renew the area has been very successful.

Urban renewal in Temple Bar

Key points

- **Urban renewal** means replacing or renovating buildings in the city.
- **Urban redevelopment** means demolishing run-down buildings. They are replaced with offices and

Planning of New Towns

Key words

Urban sprawl
Traffic congestion
New towns

Industrial tax-free zone
Export market
Administrative

Cities are growing very rapidly. Such rapid growth has caused many problems for the people who live in them.

Problems such as **urban sprawl**, **traffic congestion**, unemployment and crime are the result of such rapid growth. To deal with these problems, the government has come up with a number of projects. You have already studied an urban renewal project. Another type of project is planning and building **new towns**. Let's look at Shannon New Town.

Case Study: Shannon New Town

The government set up an army base and airport at Shannon in 1941. It was a flat site, perfect for an airport runway. It had plenty of space. Fog was not a problem. Soon 80% of flights crossing the Atlantic landed at Shannon.

Shannon Airport became the world's first duty-free zone. A duty-free zone is an area where government taxes are not paid on goods. Shannon had the first ever duty-free shop – a two-metre counter that sold alcohol and cigarettes.

Great value shopping at Shannon duty free

Shannon became the first **industrial tax-free zone**. Companies that set up within the airport zone did not pay taxes on goods made for the **export market**. This was good for businesses. Many factories were set up here. A town was planned at Shannon for the workers.

The planners began by laying out the town centre. The **administrative** offices of the town, as well as shops, would be found here. Schools and roads were added. Industrial and housing zones were planned and built. Playing fields and open spaces were an important part of the plan. This was a major attraction for people who came to live in the new town.

People lived near their work so traffic congestion was not a problem. The houses were not small and cramped together, as there was plenty of space in the newly planned town. The demand for housing and services in the nearby city of Limerick and town of Ennis decreased, as people settled in the new town of Shannon.

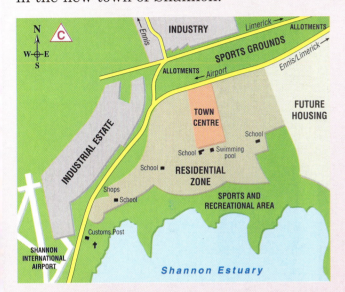

Today, Shannon is a thriving town of more than 12,000 people. It is still attracting people as it is a well-planned town, without many of the problems often found in big towns.

Case Study: Tallaght, Co. Dublin

Tallaght is a **new** or 'overspill' **town**. It was built to help solve the problem of overcrowded housing in inner-city Dublin. People were moved out to the new housing areas.

The planners agreed that the 'new town' of Tallaght would have:

- A mix of public and private housing.
- Industrial estates to provide work.
- A road system which would separate local and through traffic.
- Neighbourhood centres with churches, schools and basic shopping facilities.
- Open space for recreational purposes such as playing football.

Many young families moved out of the poor inner-city areas into the new housing estates in Tallaght. It took time for the new town to develop. Now it forms an energetic and independent community of over 100,000 people. It has:

- A town centre which has given it a heart or focal point.
- A regional third-level college and a major hospital.
- A shopping centre and a theatre and arts centre.
- Major transport links with the centre of Dublin and beyond.

Tallaght is now the centre of Ireland's newest administrative area, South Co. Dublin.

The Square at Tallaght

Key points

- Shannon (Co. Clare) and Tallaght (Co. Dublin) are planned **new towns**.
- These towns are planned to take pressure off nearby larger towns.
- Housing, education, leisure and other services are provided for locals.

REVISION EXERCISES

Write the answers in your copybook.

1 Urban renewal is when:
   ● New houses are built on open spaces in the countryside
   ● Large industrial estates are built
   ● Buildings in the city are demolished and replaced with shops
   ● Old buildings in the city are restored or replaced

2 Shannon is:
   ● A major university
   ● A planned town
   ● A redeveloped area
   ● A renewed area

3 Choose the correct answer for **each** of the statements below.
   (a) Tax breaks and grants are offered in **designated/designer** areas.
   (b) Urban redevelopment means more **houses/offices**.
   (c) Temple Bar is an area of urban renewal in **Dublin/Cork**.

4 In Western cities many planners are trying to overcome problems in run-down areas. Write out the **incorrect** words in these statements.
   (a) Rates and taxes are being **increased/decreased** as an incentive for developers.
   (b) High-rise apartments are being **encouraged/discouraged**.
   (c) Older houses are being **preserved/replaced by factories**.

5 Urban renewal projects are seen in many Irish cities. In the case of **one** city that you have studied describe the project and how it changed the area.

6 Urban redevelopment rather than urban renewal is often preferred by developers. Give **one** argument in favour of and **one** argument against redeveloping an old run-down part of the city.

7   *Urban planning can sometimes improve the environment in which people live.*
    Name **one** example of an Irish city or town that you have studied that has seen good planning. Explain **two** ways in which urban planning has improved the environment.

8   Name **two** advantages and **two** disadvantages of living in a new town compared to living in an older, more established area. Use information that you have learned in previous chapters including population pyramids.

9   Give **one** reason why business people opened their factories in Shannon.

10 Look at the picture.
    (a) Explain what is meant by the phrase 'Section 23 Tax designated areas'.
    (b) Discuss **one** way that identifies an area as a designated area.
    (c) Describe **one** advantage of investing in a property that is located in a designated area.

11 Areas that have been renewed often develop a lively social life. In the case of **one** scheme that you have studied describe why people gather there during their free time.

16 Urbanisation in Poor Countries

Key words

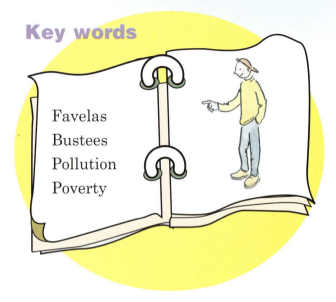

Favelas
Bustees
Pollution
Poverty

Cities in the developing world are very different to cities in the developed world. Two main differences stand out.

These cities are growing at a very fast rate. In-migration and a high birth rate mean that the cities cannot cope. Rapid population growth results in vast areas of unplanned housing at the edges of these cities. These are called **favelas** or **bustees**. They are not a feature of cities in the developed world.

Another difference is that the gap between the rich and poor in these developing world cities is far wider than in cities of the developed world. Let's look at São Paulo in Brazil which will show these differences clearly.

Case Study: São Paulo, Brazil

São Paulo is the capital of the state of São Paulo in south-eastern Brazil. It is the biggest city in Brazil. Between 18–20 million people live in the Greater São Paulo Metropolitan region. This vast city is ranked as the fourth largest in the world.

Like other cities in developing countries **São Paulo** is a city of great contrasts. On the one hand, it is like any city in the developed world. It is a modern, lively city with many concerts, plays and international sports events. It is the most important centre for business in South America.

The city cannot cope with a rapid growth in population numbers. As a result, there is poor housing, a lack of basic services and high pollution levels.

Housing

Almost 40% of São Paulo's population live in the **favelas**. Often favelas are built on land that is unsuitable for housing. The land may be steep or it may be easily flooded. Landslides often occur. Homes are made of wood or other flimsy material so the risk of fire is high. These homes are overcrowded.

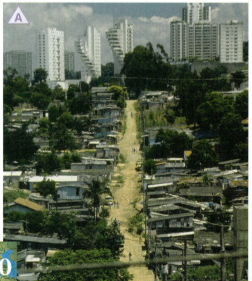

Favelas in São Paulo

The favelas are often lawless, with high levels of crime and drug use. Even the police fear going into them!

Services

A large population means that there is pressure to supply basic services. There are problems with bringing clean water to the houses in the favelas and supplying them with electricity.

A large population means that **pollution** levels are high. Rubbish collection is not guaranteed. A proper sewage system is not available to everybody. The risk of disease is high.

Pollution

Pollution of land, air and water is a major problem. Small, unregulated industries, set up within the favelas, create much of this. Unregulated means there are no controls on their activities. Pollutants from the use of chemicals used in dying clothes for example, are a health hazard.

With such high population numbers it is almost impossible to police all the activities in the favela.

Gap between rich and poor

In São Paulo the difference between the rich and poor is very obvious. This is a world of extremes. A small number of very rich people own most of the wealth. The vast majority of people are very poor. This contrasts with cities in the developed world where there is a greater spread of wealth.

The gap between rich and poor can be found in the areas of housing, healthcare and access to education. The **poverty** of the favelas is in stark contrast to the luxury apartments of the wealthy. Iron railings and security people guard the entrances to these dwellings.

The children of the wealthy attend expensive private schools. Private hospitals serve the better off. The poorer people, the vast majority, depend on an under-funded public health service.

Key points

- The gap between the rich and poor is wider in cities in the developing world than the developed world.
- In the developing world, cities cannot keep up with the demand for houses and other basic services.
- A zone of unplanned housing called a **favela** forms a ring around the edge of cities such as São Paulo.
- **Favelas** are not a feature of cities outside the developing world.

REVISION EXERCISES

Write the answers in your copybook.

1   Cities in the developing countries are most likely to have:
   ● Large expanding central business districts and new suburban business centres
   ● Increasing populations with poor housing and services
   ● Decreasing population as people move to rural areas
   ● Well-planned suburban housing estates with good services

2   Choose the correct answers from the statements below.
   (a) São Paulo is the **tenth/fourth** largest city in the world.
   (b) Favelas are usually on the **edge/centre** of cities of the developing world.
   (c) There is more planned housing in the **developed/developing** world.

3   Match each letter in column X with the number of its pair in column Y.

X	Y	ANSWER
A Favela	1 Heavy rainfall	A =
B Higher birth than death rates	2 Shanty town	B =
C South America	3 Natural increase	C =
D Landslides	4 São Paulo	D =

4   Give **two** reasons why so many people have migrated to the city of São Paulo.

5   There are many problems associated with the rapid increase of population in cities in developing countries. Discuss **two** of these problems in a city of your choice.

6   Many of the wealthier houses in the developing world have armed guards protecting them. Suggest **one** reason why that is so.

7   There is poor planning in the favelas. Explain **two** ways that this affects the lives of the people who live there.

8   Using an atlas, describe where São Paulo is located. In your
    answer refer to:
    ● The hemisphere in which it is located and at what latitude
    ● The name of a nearby ocean or river
    ● The type of coastline on which it lies – straight or sheltered
    ● The name of the nearest big city or town

9   Imagine that you are a planner based in São Paulo. You have been
    asked to prevent the spread of the favelas. Discuss **one** way that
    this may be done without costing too much money.

10 Look at the photograph of a
    favela in South America.
    (a) Describe **two** features of a
        favela which are shown in
        the photograph.
    (b) Give **one** reason why
        favelas lack the basic
        services.

11 Imagine you are visiting São Paulo. Write a letter to a friend in
    which you describe the city. Refer to the main difference you notice
    between this city and a city in Ireland.

17 Natural Resources

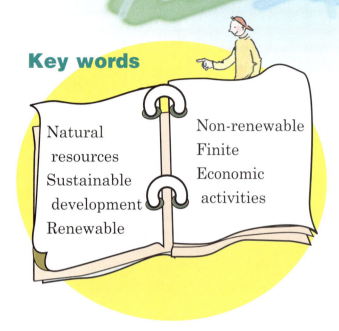

Key words

Natural resources
Sustainable development
Renewable

Non-renewable
Finite
Economic activities

A	Renewable resources	Non-renewable resources
	Soils	Coal
	Wind	Oil
	Tides	Gas
	Fish	Iron ore
	Sunshine	Gold
	Rain	
	Trees	

The earth has supplied us with the **natural resources** that are needed for our survival. These include the important resources of water, rocks and soil.

It is important to take care when using the earth's natural resources, so that we do not damage them forever. This 'good use' is called **sustainable development**. It is about making sure that the earth can continue to support or sustain life in the future.

Renewable and non-renewable resources

Some natural resources are **renewable**. This means they can be used again and again. Soil is an example of a renewable resource that with good care can be reused.

However, other natural resources are **non-renewable** or **finite**. They run out eventually. Oil is an example of a non-renewable resource.

Economic activity

Economic activities involve using the earth's natural resources. These activities can be grouped as follows:

- Primary activities
- Secondary activities
- Tertiary activities.

Think of this grouping as a way of describing particular kinds of jobs from which people earn a living.

Primary activities

Primary activities are jobs that have to do with working directly with the natural resources of the earth.

Farmers, forestry workers, miners or fishermen work in primary activities as they work directly with soil, trees, rocks or the seas.

Raw materials are the materials that come directly from the soils, rocks and waters of the earth. Examples of raw materials are crops from the soil or minerals from rocks. At a later stage raw materials will be processed (made) into other products.

Forestry is a primary activity

Secondary activities

Secondary activities are jobs that involve the manufacturing (making) of products. This work generally takes place in factories.

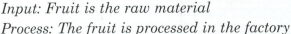

Input: Fruit is the raw material
Process: The fruit is processed in the factory
Output: Pots of jam are produced

When a farmer harvests crops they are sold on to the factory. The factory workers take this raw material of crops and they process them or use them to make other products. For example, jam is made from fruit.

In a similar way, the rocks of the earth can be ground down and turned into concrete. Metals can be melted down to make metal products. For example, iron ore can be used to make steel.

Fish from the sea can be cleaned out and tinned or packaged for sale.

At the secondary stage of work, goods already manufactured can be further used to make other products. For example, steel can be used to make cars or other machines.

Tertiary activities

Tertiary activities are jobs that provide a useful service to people. This group of workers includes shopkeepers, hairdressers, train drivers, nurses and teachers.

Nursing is a tertiary activity

Driving is a tertiary activity

Key points

- The earth has provided us with **natural resources** of rocks, soil and water.
- **Natural resource**s are **renewable** and **non-renewable**.
- The three types of **economic activity** are primary, secondary and tertiary.

Water: A Renewable Natural Resource

Key words

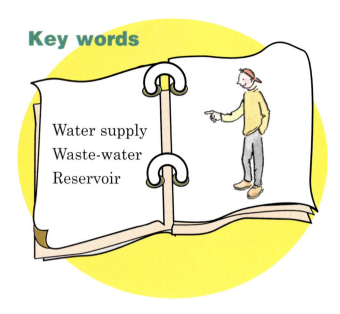

Water supply
Waste-water
Reservoir

Different ways of receiving a water supply

Water is a renewable resource. It can be used again and again.

Fresh water

Although more than 70% of the earth's surface is covered in water, most of this water is in the oceans. It is salty and is unsuitable for use in homes, industry or farms. A mere 2.6% is fresh and not all of this is available for use because it is locked up in icecaps and glaciers.

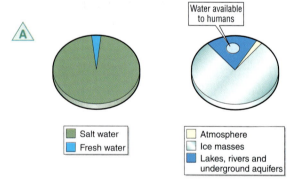

Water available to humans

Salt water
Fresh water

Atmosphere
Ice masses
Lakes, rivers and underground aquifers

Having a clean and reliable **water supply** is very important. While there is a severe shortage of clean water in many parts of the world, for most of us in Ireland it is just a case of turning on the tap or flushing the toilet. But a water supply does not arrive by magic. Supplying clean water and taking **waste-water** from homes is a big operation.

Irrigation

In some very dry parts of the world, water is carried by pipeline into the fields. This is called irrigation. By using irrigation, crops such as wheat, sugar cane and rice can be grown on land that was once too dry for crops to survive.

Let's look at a case study of a local water supply.

Case Study: The East Waterford Water Supply Scheme (EWWSS)

The EWWSS was built to supply Waterford City and the east of the county with water. The water supply comes from three sources: Knockaderry Lake, Ballyshonock Lake and the River Clodiagh. The water is pumped from the reservoir to treatment plants where it is cleaned and made fit for drinking. One of the treatment plants is at Adamstown.

This map shows the reservoirs in Co. Waterford

1 Water treatment plants

Water cannot be pumped directly to your home from a **reservoir** (storage lake). When water is taken from the reservoir it has to go through different treatment stages until it is suitable for drinking.

2 Filtering to remove impurities

Water is passed slowly through a bed of fine sand that is about one metre deep. This filters out bacteria and viruses in the water.

3 Chemicals are added

Chlorine is added to make sure that the water is clean.

Fluoride is added to the water supply to help prevent tooth decay in children.

Chemical coagulants are added so that the water has a clear colour and does not have a smell.

4 Storage

The treated water is then stored in a *covered* reservoir. The water is drawn off to supply homes, businesses and other users.

The waste-water from houses has to be treated again before it is delivered back to rivers and the sea.

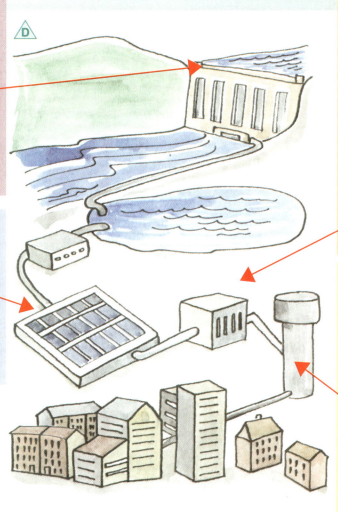

A water treatment plant

Key points

- Water is a vital natural resource.
- Water is renewable.
- Water can be stored in a **reservoir**, treated and made fit for drinking.
- An irrigation scheme makes it possible to grow crops in areas where there is a shortage of water.

Oil: A Finite Resource

Key words

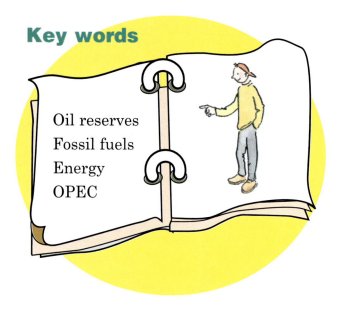

Oil reserves
Fossil fuels
Energy
OPEC

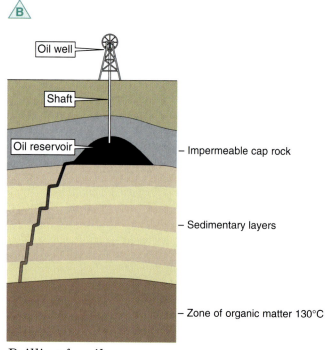

Drilling for oil

Oil is a non-renewable or finite natural resource. At the rate oil is being used, the world's **oil reserves** will be used up in less than 100 years.

Oil – a fossil fuel

Oil and gas are found under the sea floor and under landmasses. They are known as **fossil fuels** because they were formed from organic matter such as trees and plants. It took millions of years to turn organic matter into oil and gas.

Oil trade

The flow of the oil trade is shown on map A. Most of the trade in oil comes from the countries of the Middle East and is bound for North America, Europe and Japan.

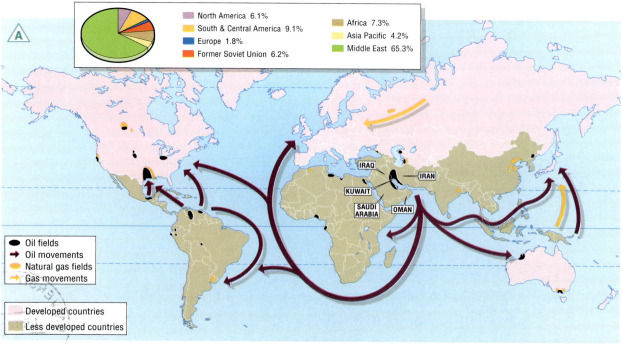

Flow of oil trade

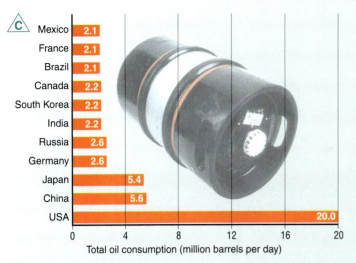

Consumers of oil

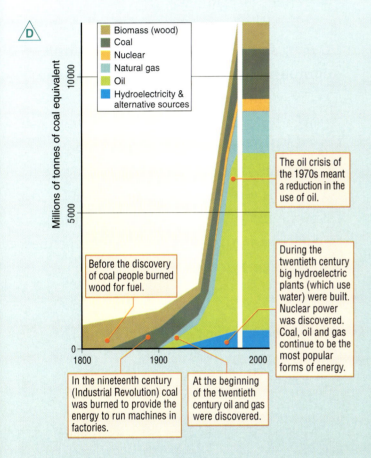

Biomass (wood)
Coal
Nuclear
Natural gas
Oil
Hydroelectricity & alternative sources

Millions of tonnes of coal equivalent

The oil crisis of the 1970s meant a reduction in the use of oil.

Before the discovery of coal people burned wood for fuel.

During the twentieth century big hydroelectric plants (which use water) were built. Nuclear power was discovered. Coal, oil and gas continue to be the most popular forms of energy.

In the nineteenth century (Industrial Revolution) coal was burned to provide the energy to run machines in factories.

At the beginning of the twentieth century oil and gas were discovered.

The 1970s oil crisis

For much of the twentieth century many developed countries became rich because there was a supply of cheap oil. In 1973, the oil-producing countries formed an organisation called OPEC. They decided to increase hugely the cost of a barrel of oil. Developed countries called this the 'oil crisis'. Countries that depended on imported oil had to cope with a massive jump in fuel prices.

Many factories couldn't pay the higher costs of fuel and wages so they closed down. Workers lost their jobs. Taxes were increased to pay for the increases in social welfare payments to the unemployed. People had to emigrate to find work.

This prompted governments to look for their own (native) energy supplies.

Changes in our use of fuels

The first fuel was wood. Later we used coal. Nowadays, oil and gas are the most popular **energy** sources. Look at the chart that shows the various sources of energy that have been used over the years.

OPEC countries – Organisation of Petroleum Exporting Countries

Saudi Arabia, Iran, Iraq, Kuwait, Oman, Quatar, Bahrain, Libya, Nigeria, Algeria, Gabon, Indonesia and Venezuela

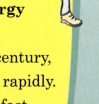

Key points

- Oil is a non-renewable **energy** source.
- Throughout the twentieth century, the use of oil has increased rapidly. **Oil reserves** are dropping fast.
- Governments have stepped up their search for native **energy** sources.

Oil and Gas in the Celtic Sea

Key words

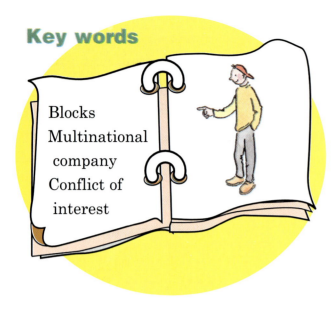

Blocks

Multinational company

Conflict of interest

Discovery of gas

Gas was discovered in the Celtic Sea off the coast of Cork in the 1980s. Since then, oil has been discovered off the coast of Waterford. Gas has been found in the Corrib field off the coast of Mayo.

Gas and oil fields, and offshore basins

Funding oil and gas operations

Searching and drilling for oil and gas is a costly and risky business. Following the oil crisis, the Irish government invited big oil companies to look for oil and gas off the coast of Ireland. They rented out sections of the sea floor, called **blocks**, to the oil companies where they could carry out their search for oil and gas.

During the 1970s a large **multinational company** (a company with operations in many countries) called Marathon began a search for oil in the Celtic Sea off the coast of counties Cork and Waterford. They discovered that there was a large gas field under the sea-bed off the Old Head of Kinsale in Co. Cork. They began drilling. Soon they were pumping gas into pipelines and bringing it ashore.

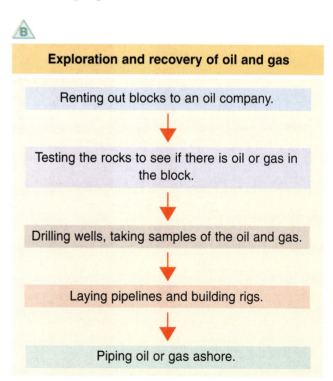

Exploration and recovery of oil and gas

Renting out blocks to an oil company.

Testing the rocks to see if there is oil or gas in the block.

Drilling wells, taking samples of the oil and gas.

Laying pipelines and building rigs.

Piping oil or gas ashore.

Oil off the Waterford coast

Oil has been discovered under the sea-bed off the coast of Waterford. Oil rigs have not yet drawn oil ashore from this block. This may happen in the future when there is an increase in the price paid for oil.

An oil rig

Oil Discovered!

Marathon Oil made $68 million in sales from the Kinsale gas field last year, and the US company now believes that the well will remain in production until 2014.

Marathon discovered the Kinsale reserves almost 30 years ago, and it became the main supplier to Bord Gáis Éireann. The field became the State's primary source of natural gas.

Following further discoveries in the region in the mid-1990s, it was believed that it could secure supply until 2015. Later reports suggested the Kinsale field would stop producing as early as 2005.

Cork-based Marathon Oil has a share in the Corrib Field and has been involved in further exploration in the Porcupine Basin off the Kerry coast and in the Celtic Sea. (From the *Irish Times* 2004, by Barry O' Halloran)

Conflict of interest

It would be a great advantage if Ireland had its own oil supply, but not everyone agrees. There is a **conflict of interest** about how this resource should be used.

Conflict of interest: not everyone agrees!

Key points

- Important oil and gas fields have been discovered in the Celtic Sea off the coast of counties Cork and Waterford.
- A **conflict of interest** could arise between those who want to bring oil or gas ashore and those who are concerned about the environment.

Energy in the Future

Key words

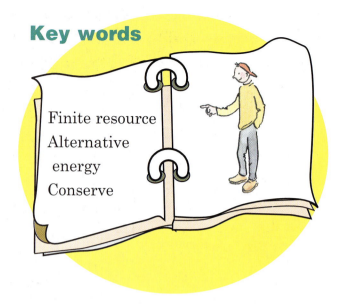

Finite resource
Alternative
energy
Conserve

Oil is a **finite resource**. It will eventually run out. The big increase in the price of oil during the 1970s gave a taste of what it could be like if there was a shortage of oil. A healthy economy depends on a reliable and affordable source of energy. For this reason it is important to do the following:

- Continue the search for oil and gas.
- Use **alternative** sources of energy.
- **Conserve** existing energy supplies.

Alternative energy supplies

Alternative energy sources include renewable energy sources such as:

- Wind power

A

- Tidal power

B

- Solar power

C

- Geothermal power

D

hot rocks heat underground water

Conservation of energy

Energy can be conserved (saved) by:

- Turning off lights or using low-energy light bulbs.
- Walking or using public transport instead of using private cars. This would cut down on the use of oil.
- Insulating attics and using double-glazed windows. This reduces the cost of heating.

Case Study of Saudi Arabia: Oil under the Desert

(This case study will not be examined at Ordinary Level.)

Oil has been described as black gold. Many countries in the Middle East have become wealthy because of the sale of their oil supplies. As you learned in the last section on oil, the price of oil rose dramatically during the early 1970s. This earned billions of dollars for the countries of the region. In this case study let's look at the changes which the discovery of oil brought to Saudi Arabia. Look at the map that shows where Saudi Arabia is located.

Saudi Arabia is in a region called the Middle East

Saudi Arabia has the world's largest oil reserves. It exports more oil than any other country in the world. It was once a very poor country. There were very low standards of healthcare and education. Its population was mainly made up of nomads who wandered around the desert with their herds of camels searching for pasture for their animals.

Saudi Arabia is now one of the world's richest countries. From the sale of their oil they have been able to make the following improvements:

● Irrigation schemes have been introduced. Many parts of the desert have been turned into fertile areas producing a range of fruits and vegetables.

● Modern cities have been built with lots of jobs for the people of Saudi Arabia.

● Hospitals and schools have been built.

● New roads have been built and industries have been set up.

Key points

● To ensure we have a reliable and affordable supply of energy we need to use **alternative energy** sources and to **conserve** energy.

● Saudi Arabia's wealth is due to its oil reserves.

Changes in Harvesting Peat

Key words

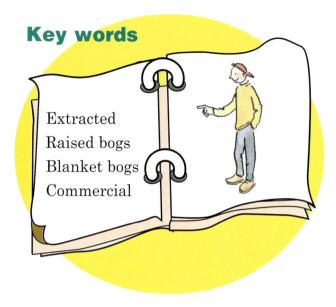

Extracted
Raised bogs
Blanket bogs
Commercial

Peat (or turf) is a non-renewable energy resource. It is used as a solid fuel in fires or is burned in power stations to make electricity. Peat can also be used as a fertiliser in the form of moss peat.

Fifty years ago peat was **extracted** (taken) from bogs mainly by hand. Its extraction was a very slow process. The use of machines has now speeded this up.

Formation of peat

Peat was formed about 10,000 years ago. Plants such as rushes and reeds piled up in lakes or badly drained areas. The layers of plant matter built up to form a bog.

Peat forming in a lake

Raised bogs and blanket bogs

You find peat on the **raised bogs** in the Midlands of Ireland and on the **blanket bogs** on mountains along the west of the country. Raised bogs are as much as 12 metres deep, while blanket bogs are less than 3 or 4 metres.

Ⓑ

Glenties

Oweninny

Mt. Dillon

Cúil na gCon

Blackwater

Ballivor

Derrygreenagh

Boora

Carrigcannon

Barna

Thick raised bogs
Thin blanket bogs

The location of peat bogs

Old methods of cutting peat

The traditional method of cutting and saving peat is shown in the following pictures. This method of cutting peat was common up until the 1950s. It did not involve machines. It was very slow.

Ⓒ

The special tool used for cutting sods of peat is called a slean

144

Stacking the peat to dry it out

People in the past would spend days on the bog helping each other to cut and save the peat for the following year. This was not a **commercial** operation. Each family cut enough for their own needs. Worked in this way, the bogs would have lasted a long time.

Bord na Móna

In the 1940s the government set up a company called Bord na Móna. They had a more commercial (business-like) approach. They wanted to speed up the whole process of cutting peat. They introduced machines onto the bogs.

Bord na Móna mainly worked on the deep raised bogs of the Midlands, where there was lots of peat for cutting. The land was fairly level on the Midlands, making it easy to use machines to cut peat. More than 50 years on, the use of machines has meant that most of the peat has been removed from the raised bogs in the Midlands of Ireland.

Using machines to cut peat involves the following steps:

- Draining the bogs by building canals and drainage ditches. Machines called ditchers are used. They do not sink down into the wet bog.
- Using machines called millers to scrape off the top few centimetres of peat.
- Leaving it to dry and then gathering it into ridges.
- Building light railways across the bog to transport the peat to the power stations.
- Transporting it to the power stations in railway wagons.

Nowadays machines are used to cut the peat

Key points

- Peat is a non-renewable energy resource.
- Machines have speeded up the cutting and harvesting of peat.
- Because of these modern machines, most of the peat has now been removed from the

Fishing: Falling Fish Stocks

Key words

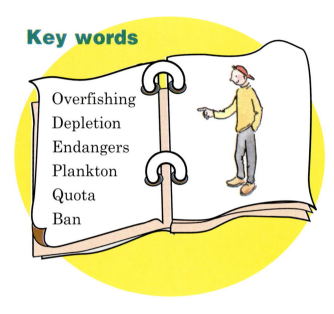

Overfishing
Depletion
Endangers
Plankton
Quota
Ban

When used carefully, the resource of fish is renewable. When misused it becomes non-renewable. **Overfishing** (taking too many fish from the sea) can lead to the **depletion** (fall) of fish stocks in certain areas.

Fall in fish stocks

Fish stocks have become run down in some areas for two main reasons.

1. Fish is a very healthy food, rich in protein and low in fat. The increase in the world's population has increased the demand for fish.
2. Modern ships and fishing equipment make it easy to catch fish in very large amounts. This **endangers** (threatens) some species of fish.

The story of catching fish

People have become really good at catching fish. Modern tracking devices such as sonar (tracking sounds) and radar equipment make it easy to locate shoals of fish.

When found, the fish are taken out of the sea with huge nets. Those that swim near the surface are caught in drift nets. Fish that swim near the sea-bed are caught in trawl nets.

When caught, the fish may be loaded onto factory ships. Here they are cut, cleaned and made ready for sale. They are sent back to the shore on other boats while the factory ships continue the business of catching and processing more fish. They can clear a whole area of fish in a short time. The fish may never renew themselves. They die out.

Fish haven't got a chance!

Case Study: Fishing in Ireland

Ireland has many natural advantages for fishing.

- It is an island and people can fish off all our coasts in unpolluted waters.
- Ireland has a shallow sea area all around the coast called the continental shelf. Sunlight can reach the sea floor and fish food called **plankton** grows here. This attracts lots of fish into these areas.
- A warm current of water called the North Atlantic Drift or Gulf Stream flows along the west coast of Ireland. This keeps the ports ice-free.
- Ireland has plenty of sheltered harbours where boats can dock safely.

Danger!

Fish stocks in Irish waters are under threat. Because Ireland is a member of the EU, we have to share the fish with other EU states (now numbering 25). In parts of the Irish fishing areas, serious overfishing almost wiped out cod and herring stocks.

The map shows the areas where fish stocks are endangered

This map shows the fishing rights zone around Ireland

Protecting fish stocks

An agreement called the Common Fisheries Policy has been drawn up by the EU. With the backing of this policy the Irish government aims to protect fish stocks.

For the most part, only Irish fishing fleets can fish in the 'Irish box'. This zone is 320 km wide.

A **quota** or limit has been set on the amount of fish that can be caught in a year.

There is a **ban** on fishing some species at certain times of the year. This is in the breeding season. During this time the fish population can renew itself.

Key points

- Modern fishing methods have led to a fall in fish stocks in parts of the world.
- **Quotas** and **bans** on fishing in certain areas at certain times of the year can protect fish stocks.

:

Farming as a System

Key words

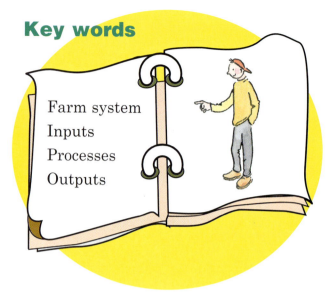

Farm system
Inputs
Processes
Outputs

Let's look at farming as a system. This means farming works in an organised way. Think about it as a series of:

- Inputs
- Processes
- Outputs.

Now picture this! You have just inherited a farm and your challenge is to make a going concern of it. What will you need to know at the start? Knowing key farming words will help!

Inputs

Inputs are all the things that are in place, or that you will have to put in place, before and while you carry out your farming activities.

- How big is your farm?
- What kind of land is it? Is it flat or hilly?
- Is the soil fertile or infertile?
- Will you have to put fertilisers on your land?
- What is the weather like where your farm is located?
- What kind of machinery will you use?
- What crops or animals will you have?
- What kind of buildings will you need?
- Will you need other workers to help with the farm work?
- How much money will you need to buy seeds, machinery, and fertilisers?

Processes

Processes on farms have to do with the jobs you will be doing on the farm. Think about processes as the various activities that are involved in farm work. You could be doing any of the following jobs.

- Ploughing fields.
- Sowing seeds.
- Reaping crops.
- Milking cows or goats.
- Building drains.
- Spraying with insecticide or fertilisers.
- Mending fences.
- Feeding animals.

Outputs

When you have done all this work you will then see the fruits of your labours! You may not have all of the following **outputs** but you might have some of them.

- You may have milk to take to the creamery.
- You may have cattle to take to market.
- You may have eggs to sell.
- You may have wool to sell.
- You may produce silage or hay to feed to your animals.
- You may have manure to spread on your fields.

Ploughing fields is part of the process

Taking milk to the creamery

Key points

- Farming is a system of **inputs**, **processes** and **outputs**.
- **Inputs** include soil, climate, farm machinery and fertilisers.
- **Processes** are the jobs done on farms. These include ploughing the land or mending fences.
- **Outputs** include crops, meat, milk or manure.

Farming in Ireland

Key words

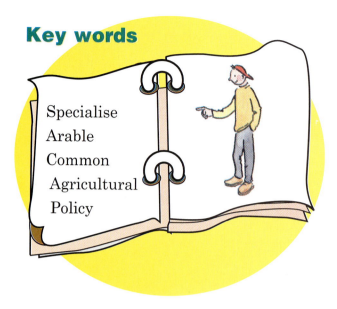

Specialise
Arable
Common
Agricultural
Policy

Farming is an important part of the Irish economy. It employs about 6% of the Irish workforce directly. If you add to that figure the number of people who work in related industries such as transport, food processing or fertiliser production, you find that many people depend directly or indirectly on farming for their living.

Membership of the EU has had a great influence on farming. The **Common Agricultural Policy** (CAP) influences prices that farmers can expect for their produce. It has an influence on decisions about what to do on the farm.

Case Study: a Mixed Farm

A farmer can decide to **specialise** in one activity on the farm. It might be dairy farming or growing wheat. Or he/she may decide to do a mixture of things. This is called mixed farming.

On a mixed farm, a farmer will rear animals and grow crops. His or her activities will be a mixture of **arable** and pasture farming.

Arable farming involves ploughing land. Pasture involves letting animals graze on grass or feed on silage or hay.

Picture yourself as a farmer on a mixed farm. This is your farm.

You have 400 cattle on your farm and you rear some of these for their milk and sell some on to other farmers for fattening. You grow the crops wheat and barley and a small amount of vegetables. You have a milking parlour. You have a barn in which your animals stay in the winter.
You have a silage area where you store the winter feed for your animals.

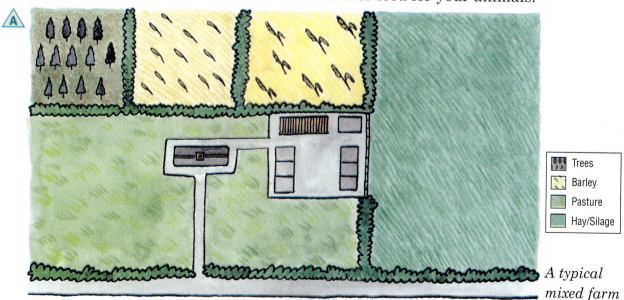

Trees
Barley
Pasture
Hay/Silage

A typical mixed farm

This would be the work programme for the year.

B	Yearly cycle of work	
Season	Type of work (processes on farms)	
Winter	Stall feeding of cattle Mending fences Repairing buildings	
Spring	Sowing crops Milking cows Spreading manure in fields	
Summer	Cutting silage Milking cows Spraying crops	
Autumn	Harvesting crops Ploughing the land Taking cattle to market	

Labour intensive work

The yearly programme of work shows that farming is a demanding job. Farmers are busy all year around. This kind of job is called labour intensive. When farmers are not directly involved in farm work they could be updating their skills by attending lectures on the latest improvements in farming or doing their accounts.

It is a job that has changed a lot in recent years. It is a business rather than a living.

Quotas laid down by the EU mean we have to limit the amount of milk produced.

Hygiene laws are very strict now.

C

Key points

● A mixed farm is one where there is a mix of animal and crop farming.

REVISION EXERCISES

Write the answers in your copybook.

1   Which one of the following is a renewable resource?
- Peat
- Solar power
- Coal
- Oil

2   Explain the following terms:
- Sustainable development
- Renewable resource
- Non-renewable resource.

3   Match each letter in column X with the number of its pair in column Y.

X	Y	ANSWER
A Primary	1 Nurse	A =
B Secondary	2 Stores water	B =
C Tertiary	3 Farmer	C =
D Reservoir	4 Baker	D =

4   Which **one** of these is a tertiary activity?
- Mining
- Fishing
- Teaching
- Making computers

5   In the case of a water supply that you have studied, describe the stages involved in bringing the water from its source to a nearby city.

6   In your copybook show how oil is formed. Use the following guidelines: sedimentary rock layers; an oil platform; drilling equipment; pipelines.

7   Name **four** countries that supply oil and describe **two** benefits that an oil find may bring to an area.

8   Give **one** reason why locals might welcome an oil find off the Waterford coastline and **one** reason why they might object to bringing this oil ashore.

9   Peat has been a major resource in Ireland. Describe **two** ways in which we have exploited (made use of) that resource in Ireland.

10 Explain why most of the peat (turf) cutting by Bord na Móna has been on the raised bogs of the Midlands.

11 The rate at which the raised bogs of the Midlands have been exploited has speeded up. Explain, using two examples of machines, how this has happened.

12 Describe **two** results of overfishing in the seas around Ireland.

13 *The number of fish in Irish waters has been reduced in recent years.* Describe **two** steps that are being taken to limit the effects of over-fishing.

14 A mixed farm has animals and crops. Describe **two** inputs and **two** outputs of a mixed farm.

15 Look at the pie chart of the use of farmland in Ireland. What is 20% (one fifth) of the land used for?
● Pasture
● Hay
● Cereals
● Other crops

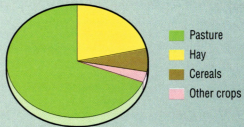

16 Describe the work that is done on a mixed farm during the year i.e. in spring, summer, autumn and winter.

18 Secondary Industry

Key words

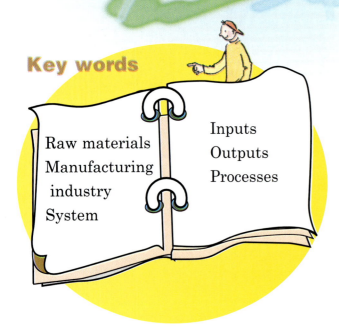

Raw materials
Manufacturing industry
System

Inputs
Outputs
Processes

Secondary industry is the stage at which **raw materials** are changed into finished products. This usually takes place in factories. This type of activity is called **manufacturing industry**.

Raw materials are the ingredients from which things are made. These raw materials come from farms, forests, mines and the sea. They are described as raw because they are just as nature made them.

More key words!

Key words to do with manufacturing:
Industrial location: Where you would find a factory.
Factory site: The actual ground on which the factory is built.
Multinational companies (MNCs): Large companies that are set up in more than one country.
Light industries: Make products which are light in weight e.g. watches and computers.

Heavy industries: Make products which are heavy in weight e.g. ships or planes.
Footloose industries: Industries which are not tied down to a particular location. There is a choice about where the factory is set up.

Factories as systems

A factory works like a **system**. A system is an activity where **inputs** are *processed* and turned into **outputs**.

- **Inputs** include the raw materials needed to make something. They also include the labour, buildings and capital (money).
- **Processes** are activities within the factory. This is usually the making of goods. It can also be design and research.
- **Outputs** leave the factory. This can include the finished product, profit or even waste.

The factory system: stages in the making of paper

Case Study: Cadbury Chocolate – a System

Let's look at Cadbury chocolate as an example of a system.

Inputs	Processes	Outputs
Cocoa beans	Sorting	Chocolates
Chocolate	Cleaning	Cakes
Fresh milk	Roasting	
Electricity	Mixing	
Labour	Drying	
Machinery	Packing	
Paper		

The making of chocolate

Step 1. Cocoa beans are delivered to the factory in Coolock, Dublin from Ghana, The Ivory Coast, Cameroon and Brazil.

This map shows where cocoa beans come from

Step 2. The beans are sorted and cleaned. They are roasted, kibbled (broken into smaller pieces) and winnowed (shell is removed from the bean). This produces cocoa mass, which contains cocoa butter.

Step 3. This cocoa mass is then transported to Rathmore, Co. Kerry. It is mixed with fresh milk and sugar and this mixture is then cooked slowly. It is dried to form chocolate crumb. It is then sent back to Dublin.

Step 4. The crumb is finely ground between enormous rollers. Extra cocoa butter and special chocolate flavourings are added to what is now a paste. The paste is beaten to develop the final flavour and smoothness.

Step 5. The paste is cooled carefully so that it is smooth. The bars are shaped, wrapped and packed for the market.

Step 6. The products are transported to the shops where they are sold.

Key points

- **Manufacturing industry** involves making products.
- The factory works like a **system**. There are **inputs, processes** and **outputs**.
- The manufacture of chocolate by Cadbury Ireland is an example of a **system**.

Location of Industry

Key words

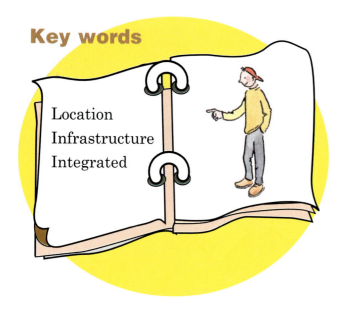

Location
Infrastructure
Integrated

Imagine that you are a factory owner. What would you have to consider when thinking about where to set up your business? Your aim would be to make a profit. There are many factors that may guide you in your choice of **location**. Let's look at five factors from diagram A.

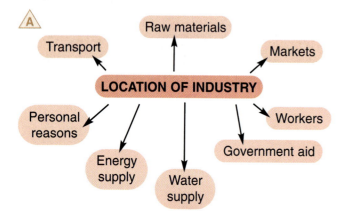

Transport and infrastructure

Raw materials, labour and finished products have to be moved in and out of factories. They need good **infrastructure** to do this. Infrastructure means the transport network that serves the factory.

If transport is poor, there may be high costs. Poor links will cost time and in business 'time is money'. An **integrated**

transport system is needed in many cases. Integrated means that the separate transport links are well-linked.

Integrated transport links

Raw materials

The factory owner has to consider whether to locate the factory near the source of the raw material or near the market.

If the factory uses food raw materials, it would be useful to set up close by. In the case of a factory making apple juice, it would make sense to locate close to orchards because the fruit would be fresher if transported over a short distance. Apples are bulky, so it is more costly to transport them at the raw material stage than at the finished product stage.

Markets

Some factories need to locate close to their markets. Once again, this is linked to transport costs.

Let's look at the brewing industry i.e. the making of beer. Here, a small amount of raw materials is added to water. It is then bottled to make a much heavier and more fragile product. For these factories it makes sense to locate near their market.

Markets, especially big ones like the EU with a population of over 350 million, attract industries. Many industries, in particular the computer industry, have been set up in Ireland because it is close to the EU market.

As a member of the EU, Ireland can sell goods in Europe without import taxes. American companies, by setting up within the EU zone can benefit from this.

Ireland has a well-educated workforce

Government aid

In recent times, government aid has played a big role in attracting multinational companies to Ireland. The government encourages companies by giving tax breaks and grants to the owners to locate here.

Setting up in Ireland, a member of the EU, is a good idea

Educated and skilled workforce

An area or city with many young people, universities and advanced colleges is attractive for industries. Many foreign companies have set up in Ireland because we have a well-educated, highly-skilled workforce.

Government aid encourages companies to locate to Ireland

Key points

- Factory owners want to cut costs and improve profits.
- To achieve this a good **location** is important.
- When choosing a **location**, factory owners consider transport, raw materials, markets, workers and government aid.

Case Study: A Local Factory

Key words

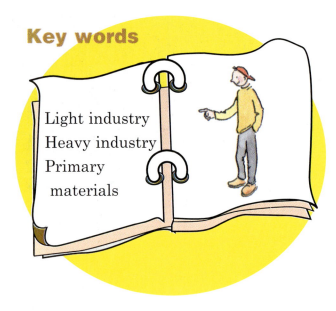

Light industry
Heavy industry
Primary materials

The manufacture of chocolate is a **light industry**. Light industries do not need a lot of space. Light industries use lightweight, raw materials and produce lightweight goods. They can be near large housing estates where workers live.

Let's look at Cadbury's chocolate factory as an example of a both a local factory and a light industry.

Cadbury Ireland in Coolock, Dublin employs over 1,700 people. Another 6,000 are employed in providing milk, sugar and packaging to the factory.

Reasons for location in Coolock

- There was a large site available in Coolock.
- Coolock is close to a large, skilled labour force. Many of the workers live in the local area.
- The site is close to excellent transport networks. The M50 motorway is close by. Goods vehicles can by-pass the city centre and avoid traffic congestion. It is near Ireland's main cargo port at North Wall. Dublin's International Airport is nearby. Cadbury exports chocolate products to the UK, USA, Canada and Japan.
- With the support of both the Irish government and the Industrial Development Authority, the factory has increased the number of job opportunities for local people.

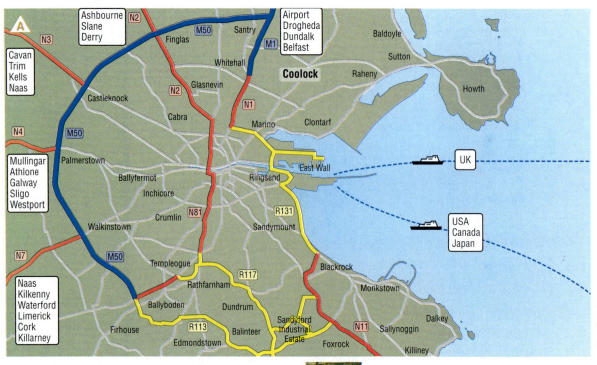

The location of Coolock in Dublin

Cement Roadstone – a heavy industry
(Higher Level only)

Cement Roadstone is an example of a **heavy industry**. The company uses bulky, heavy raw materials to make bulky, heavy goods.

Roadstone employs over 1,500 people at more than 50 locations in Ireland. It is one of Ireland's largest industrial companies with branches in mainland Europe, the Americas and the UK.

The company was involved in the building of Dublin's East Link and West Link toll bridges. It has contracts for surfacing many of our roads.

Reasons for setting up at Belgard, Co. Dublin

- Raw materials, such as limestone rock and water, are available in the nearby quarry and from local rivers.

Roadstone factory

Cement Roadstone at Belgard in Dublin

The factory at Belgard produces crushed rock and finished materials such as concrete blocks.

Production began in the local quarry in 1970. The quarry floor has now reached a depth of 45 metres. There is enough rock for production to last another 50 years. The entire process is highly automated (machines do a lot of the work).

- The site is a nodal point. There are transport links from here to the rest of the country and abroad. The factory is near the N7 and the M50.

- The local area provides an excellent workforce.

- A large site was required as this factory uses heavy machines and needs plenty of parking space for its heavy vehicles. When the rock face is blasted, up to 12,000 tonnes of rock is released. Space is needed for this process to be safe and efficient.

Key points

- Cadbury Ireland is an example of a **light industry**.
- Cement Roadstone is an example of a **heavy industry**.
- Raw materials, transport, workforce and space are factors that influence the decision to locate to an area.

Modern Industry: A Footloose Industry

Key words

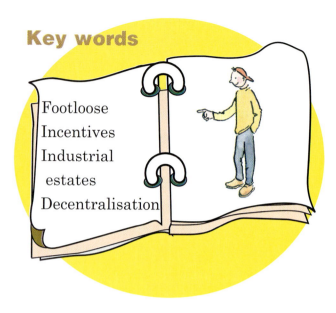

Footloose
Incentives
Industrial estates
Decentralisation

Modern industries, such as the computer industry, tend to be **footloose**. They are not tied to a particular type of location. These factories don't need to be near the market or close to the source of the raw materials. This frees them up and allows them a greater choice about where to set up. Governments have a huge influence on this. They encourage companies by offering **incentives** such as tax breaks to locate to a particular place.

Intel produces computer chips and communication machines. It is a modern footloose industry

Case Study: Intel Ireland

Intel Ireland is an example of a modern footloose industry. This large, American-owned computer company chose a site at Collinstown Industrial Park, Leixlip, Co. Kildare for its factory.

- Modern industries such as Intel can take advantage of cheap land outside cities. They are not tied to the centre of town.
- Government tax incentives and grants are often given to companies to encourage them to locate to areas outside cities.
- Intel is a multinational company. They set up in Leixlip, Co. Kildare because of easy access to the EU market.
- Computer factories use lightweight parts. They have lower transport costs than heavy industries. They are not tied to the ports. Intel can afford to locate at a distance from a large city like Dublin.
- Over 4,700 people work in Intel in well-paid jobs. They can afford the high cost of car transport. This allows companies to locate outside large cities.

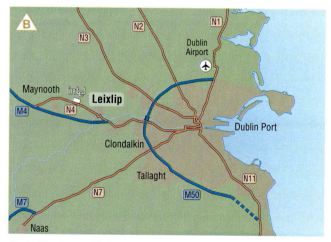

Leixlip is well located close to good transport links

Industrial estates

New modern factories have been set up on **industrial estates**. In recent times, most towns in Ireland have developed these planned estates for factories. These are sites, usually on the edge of town and cities, which have many factories. The Shannon Industrial Estate was the first such estate in Ireland.

Industrial estates attract factories because:

- They have good transport links.
- They are near housing estates where there are plenty of workers.

- They have well-serviced sites with electricity supplies, telephone lines and plenty of parking spaces.
- They have plenty of open space needed for factories which use heavy machines.
- Factories on the estate can supply each other with raw materials.
- They are encouraged by government grants to locate away from the main towns as part of the **decentralisation** plan.

Shannon Industrial Estate

- Excellent transport: Shannon International Airport.
- Government support: Tax-free status.
- Skilled labour force: Limerick, Ennis and Shannon town.
- Big site: 243 hectares.
- Planned and serviced space for other factories.

Shannon Industrial Estate

Key points

- Modern industries tend to be **footloose**.
- Governments have an important influence on where modern industries are set up.
- Ireland has attracted modern industries with its well-educated workforce and location in the EU.
- An **industrial estate** is an estate of factories with modern services.

The British Iron and Steel Industry: A Change of Location

(This topic will be examined at Higher Level only)

Key words

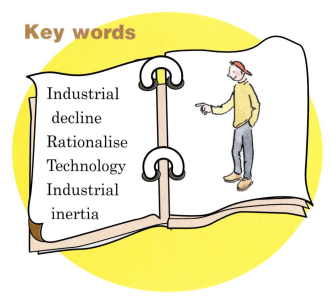

Industrial decline
Rationalise
Technology
Industrial inertia

A

1. Fifeshire
2. Central
3. Midlothian
4. Ayrshire
5. Northumberland & Durham
6. Cumberland
7. Lancashire
8. Yorkshire
9. Derbyshire
10. Nottinghamshire
11. Leicestershire
12. Warwickshire
13. North Wales
14. North Staffs
15. Shropshire
16. South Staffs
17. South Wales
18. Forest of Dean
19. Bristol & North Somerset
20. Kent

Coalfields in Britain

The British iron and steel industry has changed its location over the years.

- At first, charcoal (made from wood) was needed to smelt the iron out of rocks. Iron factories were set up close to forests such as the Forest of Dean.

- Then coke (from coal) was used to smelt out the iron. In the eighteenth century, the iron factories were set up near coalfields. Coalfields such as in South Wales often had an extra attraction: iron was available there too.

- During the nineteenth century, steel was manufactured from iron. Supplies of iron ore began to run out. It had to be imported. It was cheaper to move the industry to the coast.

- In the second half of the twentieth century, demand for steel fell. Several iron and steel plants closed. Those that remained open were near ports or were in modern planned factories such as at Port Talbot in Wales.

Industrial decline

When factories close with the loss of many jobs, we call this **industrial decline**. Let's look at the factors that lead to **industrial decline**.

B

The iron and steel factory has closed with the loss of 200 jobs.

Factory closure

1. Raw materials
When raw materials run out, the industry may close down or move to another place.

3. Fall in demand
A fall in demand for a product or service means a loss of profits. The industry needs to **rationalise** or close to save money or stop it going into debt.

INDUSTRIAL DECLINE

2. Rationalisation
Cost-cutting measures may mean that workers are 'laid off'. Multinational companies often relocate to developing countries where wages are lower. Relocate means to set up somewhere else.

4. New technology
New **technology**, often using computers, may mean replacing people with machines. In these cases people lose their jobs.

Industrial inertia

Sometimes a factory stays in its original location even though it may be more profitable to move. When this happens we call this **industrial inertia**.

For example, many of the heavy steel-making factories were set up near coalfields. Now oil or gas is preferred as a source of power. In many cases, it made sense to move to a coastal location where the big oil tankers come in. Yet some factories didn't move. One such industry is the manufacture of cutlery at Sheffield in England. Some of the reasons why it has not moved are given below.

- The name of the product is linked to a particular place e.g. Sheffield cutlery.
- They have a long tradition of skills built up in the area.
- They have a loyalty to the people whose families have worked in the industry for generations.
- The government gives grants to prevent unemployment rising in an already depressed area.

Key points

- The British iron and steel industry is an example of an industry that was in **decline**.
- Cost-cutting measures result in industries closing or setting up elsewhere.
- When a factory remains in an area in spite of changes, this is called **industrial inertia**.

The Changing Role of Women in the Irish Workforce

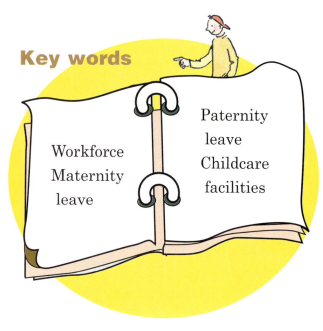

Key words

Workforce
Maternity
leave

Paternity
leave
Childcare
facilities

The role of women in the Irish **workforce** is changing.

● More and more women in Ireland are choosing to work outside the home.

● There is an increase in the percentage of older women in the workplace.

Let's look at table **A**. It shows how more Irish women are entering the workplace.

A	Percentage change in the number of women in the Irish workplace between 1971–2001	

Year	% in total workforce
1971	28
1997	47
2001	50

Let's look at the percentages in the different age groups.

B	Women in the Irish workforce	

Age	% of total females of that age in the workforce
25–34	77
20–24	76
35–44	64

Let's look at the figure for the 35–44 age group in more detail. It shows that 64% now work outside the home. This is where the real change is happening. The rate of participation of this age group in the workplace has doubled in the last ten years.

Married women in the workplace

Traditionally, when women married, they gave up their jobs. This is not happening now. The figures given in chart **C** show this change.

C	Percentage of married women in the Irish workplace	

Year	% of married women in the workplace
1991	25.7
2001	46.4

Taken together, the tables show that women are now playing a larger role in the Irish workforce. The figures also suggest that this trend will continue in future years.

Reasons for change

● Women are staying longer in education. Education improves the chances of employment.

● Women are having fewer children, making it easier to work outside the home.

● High mortgages or rents often require two wages.

● The economy needs women as well as men.

Measures to attract, support and keep women in the workplace have been introduced. These include:

- Improved **maternity** and **paternity leave** from work.
- Job-sharing opportunities.
- Improved **childcare facilities**.

Which jobs do women do in the Irish workplace?

D Jobs women do in the Irish workplace	
Occupation	**% who are female**
Managers and administrators	29
Clerical/secretarial	76
Craft and related industries	6
Plant/machine operatives	25

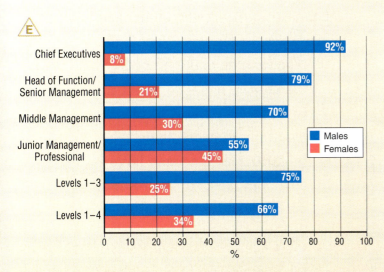

The chart shows that women tend to remain at junior management level rather than take higher positions.

Why fewer women than men reach top positions in firms

Key points

- The percentage of women in the Irish workforce is growing.
- Measures have been put in place to make it easier for mothers to work outside the home.
- Fewer women than men hold senior positions in the workplace.

Women in the Developing World

Key words

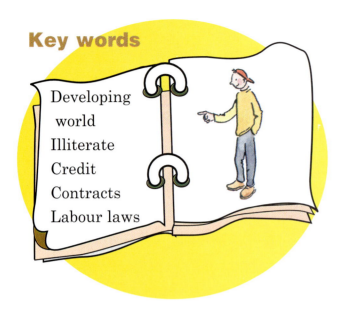

Developing world
Illiterate
Credit
Contracts
Labour laws

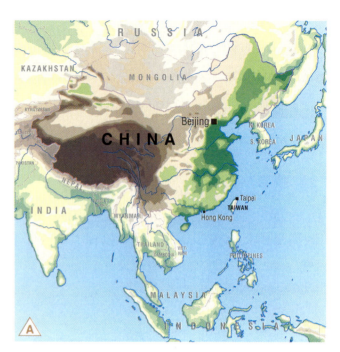

A

Let's look at some facts about women in the **developing world**.

- More than 1 billion people in the developing world live on less than $1 a day. Of these, 70% are women.
- Two-thirds of the world's 1 billion **illiterate** (unable to read) adults are women.
- Women in Africa produce 80% of the food but receive only 10% of the **credit** (loans) made available for agriculture.

It is clear that up to now the important role of women in the developing world has rarely been recognised. However, things are changing. Let's look at China.

Case Study: The Changing Role of Women in Chinese Industry

China is the third largest country in the world. It is 117 times the size of Ireland with a population of over 1 billion people.

China's economy is developing very rapidly at present. Without the contribution of women this would not be happening.

From the figures below, it is clear that in China the role of women in the industrial workforce is changing. The main ways are:

- Women now make up nearly half of the workforce.
- In cities they are working in jobs that traditionally were 'male' jobs.
- In rural areas, women have set up and now manage a range of handicraft industries.

B Women in the Chinese industrial workforce		
Year	Number of working women	% of the total labour force
1949	600,000	7.5
1978	31.28 million	32.9
2003	330 million	46.7

Low wages for women

The following statistics on wages highlight an important difference between men and women in the workplace. When it comes to wages men earn more than women. (The yuan is the monetary unit of China.)

Differences in the earnings of men and women in China		
Wage (Yuan)	% of females	% of males
2000–3000	37.2	62.8
3001–5000	32.2	67.8
5000 +	14.4	85.0

China Women, 31 March 2003

In the low-status and low-skill industries, such as handbag factories, women outnumber men by 50 to 1. Women in China and the Far East make many of the designer clothes in your wardrobe or the accessories you wear. The women who make them are paid very low wages.

These types of products are often made in China and the Far East

Changes in China – women in towns and cities

Women make up 30% of all migrants to cities. In the cities, women are now doing what were more usually considered male jobs. Women make up:

- Over one-third of workers in trade, industry, finance and communication.
- Almost half of those in education, culture and health.

Women in rural areas

Changes in the role of women are also happening in rural areas. Because 70% of the migrants from rural areas are men, women are often left at home as the main breadwinners. In these areas, women's groups have set up local craft industries to help support themselves.

Changes needed

While improvements for women in the workplace have taken place, further changes are needed.

- Women do not have equal access to education. More than 70% of China's 220 million semi-literate or illiterate people are women.
- Country people in the remote mountainous regions of China often are too poor to send their daughters to school.
- Some universities openly discriminate against female students.
- Many women work without **contracts**. They have few rights despite **labour laws** protecting women at work.

Key points

- In China, the number of women working in industry has increased greatly.
- Women earn far less than men in the Chinese workplace.
- Education and the enforcement of **labour laws** will bring further changes for women.

Classification of Industrial Regions

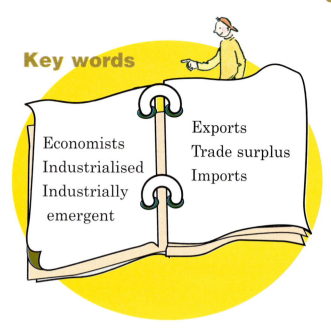

Key words

Economists
Industrialised
Industrially emergent

Exports
Trade surplus
Imports

The Industrial Revolution brought new skills, machines and products. Factories were set up and people moved into towns. But these changes did not happen in all countries at the same time. Some countries started on the road to industrial development before others did. Depending on how much industrial development has taken place, **economists** describe countries as:

- **Industrialised**
- Newly industrialised
- **Industrially emergent**.

Industrialised regions

Industrialised regions have had industries for a long time. They have built up skills and their own markets. They have also built up a reputation for good-quality manufactured goods. These are the developed regions of the world. These regions are mainly found north of the Tropic of Cancer (23.5°N of the equator). Japan, the USA and Europe belong to this group.

Newly industrialised regions

Newly industrialised regions are countries where the percentage of the workforce in manufacturing industry is growing rapidly. In these countries, a growing number of people are gaining access to education and training skills. Brazil, China and India are among this group.

Industrially emergent regions

Industrially emergent regions are countries that have been slow to become industrialised. These are the developing or poor countries of the world. They still

A

- Industrialised regions
- Newly industrialised regions
- Industrially emergent regions

depend on primary activities. This is rooted in their history. These were the countries that were colonised by European powers. They had no chance to develop industry. They supplied raw materials for the benefit of their ruling powers. Most of these countries are found south of the equator. Mali, Peru and the Philippines belong to this group.

Case Study: China a newly industrialised region

Look around you. Think of what you have bought lately. There is a good chance that it was made in China. China now **exports** twice as many goods as it did in 1996, and if recent trends continue, China will overtake Japan in the next couple of years to become the third largest exporter of goods in the world.

It is an example of a newly industrialised country. The chart below shows the dramatic increase in China's export earnings between 1993 and 2003. China is also one of the few countries in the world that has a **trade surplus**. In other words it **exports** more than it **imports**.

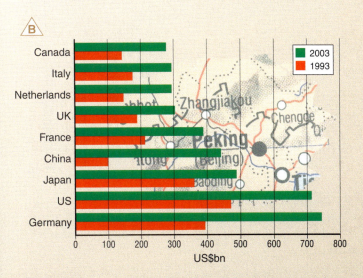

The rapid rise in China's export earnings stands out

Made in China

Between 2002–03, China showed a 17% increase in its output from factories. Workers in towns and cities now had more money to buy goods such as cars. Between 2002–03, 2 million cars were sold, a 93% increase. This shows that China is becoming a more industrialised country.

Why locate to China?

Multinational firms, such as Carrefour and Proctor & Gamble, are choosing to locate to China mainly because of:
- The improved infrastructure
- The low labour costs
- The plentiful supply of workers
- The growth of a middle class who can afford to buy more goods.

Key points

- There are different levels of industrial development in countries throughout the world.
- China is **emerging** as one of the economic wonders of the world.
- China is now the fourth-largest exporter of goods in the world.

REVISION EXERCISES

Write the answers in your copybook.

1 Manufacturing industry is:
- A secondary activity
- A tertiary activity
- A primary activity
- A service activity

2 An example of a footloose industry is:
- A computer factory
- A steel factory
- A coal mine
- An oil rig

3 Match each letter in column X with the number of its pair in column Y.

X	Y	ANSWER
A Industrial inertia	1 Systems	A =
B Footloose industry	2 Burning of fossil fuels	B =
C Acid rain	3 Not tied to one location	C =
D Inputs, process, outputs	4 An industry that is stuck in one location	D =

4 Which sentence best describes the information in the bar chart?

(a) Women in factories earned more per hour than men between 1997 and 2002.

(b) Men and women in factories have earned an equal hourly wage between 1997 and 2002.

(c) The hourly wage for women in Irish factories has increased slightly between 1997 and 2002 but is still not level with the hourly wage of men.

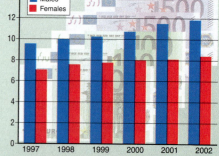

5 Using a named example of a manufacturing industry that you have studied, explain how the following factors have influenced the location of that industry:
- Transport
- Labour
- Access to raw materials

6 Describe **one** difference between light industries and heavy industries.

7 What is an industrial estate? Describe **one** advantage of building a factory on an industrial estate. In your answer include a named example of an industrial estate that you have studied.

8 Give **two** reasons why Ireland is an attractive location for multinational companies like Intel.

9 Give **one** reason why the British iron and steel industry went into decline.

10 People associate Sheffield in England with cutlery, yet this industry is said to suffer from **industrial inertia**. Explain what that phrase means. Why does the industry not move from Sheffield? Give **one** reason.

11 Describe **two** ways in which the role of Irish women in industry is seen to be different to that of men.

12 Explain, giving **two** reasons, why fewer Irish women than men reach high positions in the industrial workforce.

13 The table shows the number of female and male leaders and members of research boards and state boards in 2000. Examine the information carefully.

(a) On which board are women most represented?

(b) On which board are women least represented?

(c) Give **one** reason why men usually outnumber women in these positions.

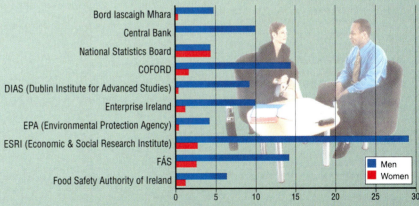

14 Give **two** detailed reasons why many factories are now moving to Asian countries like China.

19 Tertiary Activities: Focus on Tourism

Key words

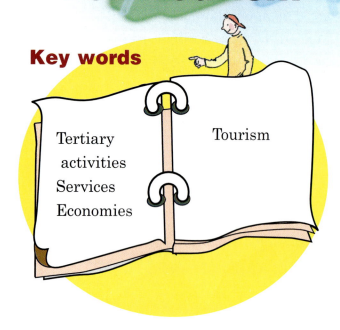

Tertiary activities
Services
Economies

Tourism

These people work in service industries

People who work in **tertiary activities** work in jobs where a service is provided. Service jobs include jobs such as teachers, nurses, drivers, shop assistants and hotel staff.

Tourism: a service industry

Tourism is the world's leading industry and is growing faster than any other industry.

Stages of development

High tax payments mean that a country can afford to supply services such as education and health **services**. High wages mean that people can afford to pay for services. For these reasons, services are more important in the **economies** of rich countries. Poor countries, for example Mali in Africa, have fewer services than rich countries.

Let's look at the following facts on tourism:

- There are over 625 million international tourists in the world each year. This is expected to rise to 1.6 billion in 2020.
- Tourists currently spend over $445 billion. This figure is expected to treble in the next 20 years.
- A country earns money from tourism. It sells beautiful scenery, good services or simply a warm welcome. These are called invisible exports.
- Tourism makes up 8% of the world's earnings from exports.

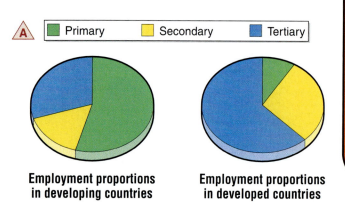

A | ■ Primary ■ Secondary ■ Tertiary

Employment proportions in developing countries

Employment proportions in developed countries

Reasons for the increase in tourism

The increase in the number of tourists worldwide has happened for a number of reasons.

- People in developed countries now enjoy paid holidays and a shorter working week.
- People are living longer. There is time to enjoy pensions and extra money when mortgages are paid off.

Look at chart E. It shows the increase in the percentage of people in the world over 60 years of age.

E	1900	1950	2000	2050	2100
	6.9%	8.1%	10%	22.1%	28.1%

- Improvements in all types of communications, such as transport, advertising and electronic links, makes travelling easier.

Impact of tourism on tourist destinations

Tourism is good for the economy, but there are drawbacks to tourism.

Positive effects
- Jobs for locals.
- Higher wages.
- Improved roads and services benefit locals and tourists.
- Greater mix of people which improves understanding.

Negative effects
- Higher costs for locals.
- Overcrowding during peak seasons.
- Seasonal jobs only.
- Increase in pollution.

Difficulties for the industry

While tourism makes a huge contribution to the economies of places, it is an industry which is not reliable. There can be good years and bad years for tourism. This creates difficulties for those countries or areas that rely heavily on it.

- The threat of terrorist attacks or natural disasters may stop visitors going to certain areas.
- Costs can rise when the price of oil increases.
- The value of a currency may change from time to time.

Key points

- **Tertiary activities** are jobs in **services**.
- Rich countries employ a higher percentage of workers in **services** than poor countries.
- The number of tourists is increasing throughout the world.
- **Tourism** has both good and bad effects.

Ireland: Tourist Regions and Facilities

Key words

Tourism
Tourist
facilities
Amenities
Regions

Tourism is very important to Ireland. Foreign visitors brought in over €4.1 billion in 2003 (diagram **A**). Almost another €1 billion was spent by Irish people themselves. This money supports people in jobs and helps to pay for **tourist facilities**.

Tourist facilities are the services that are provided for tourists. These are **amenities** such as hotels, bars, restaurants, shops, information centres, tour guides, coaches, interpretive centres, golf courses, aquatic centres and so on.

A

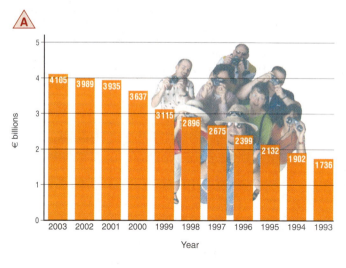

Year	€ billions
2003	4105
2002	3989
2001	3935
2000	3637
1999	3115
1998	2896
1997	2675
1996	2399
1995	2132
1994	1902
1993	1736

Employment in tourism

Over 230,000 people in Ireland are employed directly in the tourist industry. In some remote areas where there are few factory jobs available, tourism is very important. People find jobs in providing services for tourists. This cuts down migration from these areas.

Bord Fáilte

Bord Fáilte is the Irish Tourist Board. It is the main organisation for managing the tourist industry. They carry out surveys and research on tourism. They promote Ireland abroad. They provide information on tourist attractions and accommodation in Ireland.

Ireland's tourist regions

B

| Scenic areas |
| Tourist resorts |

Regions in Ireland are noted for a particular kind of tourist attraction. You will find that many tourist facilities are available in these regions. These regions are:

- Cities
- Areas of natural beauty
- Beaches and coastlines
- Recreational and sporting facilities.

Case Study: Galway City – a City Break

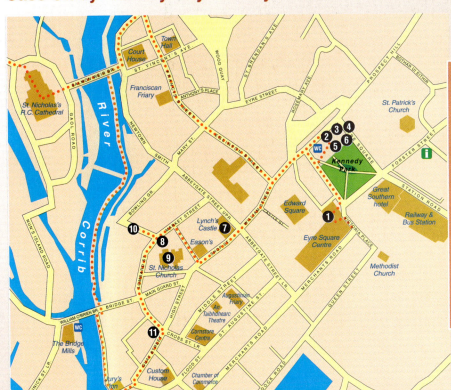

1 Medieval wall – Eyre Square Centre
2 JF Kennedy Bust
3 O'Conaire Station
4 Cannon
5 Browne Doorway
6 Quincentennial Fountain
7 Lynch's Castle
8 Market
9 St Nicholas' Church
10 Home of Nora Barnacle
11 Townhouse of Richard Martyn

Read the tourist brochure below and learn about the attractions of one city and the facilities that are there for tourists.

The city of Galway is an unforgettable city break destination

This medieval city is famously known as the 'City of the Tribes', a reference to the fourteen families who controlled the commercial life during medieval times. Come and see the remains of its medieval past. Enjoy live music and street theatre, a tradition for which Galway is famous.

Visit the Galway Atlantaquaria, the national aquarium. Here you will find an exciting display of natural habitats, from the sea-bed to local rivers and lakes.

Children will love the 'touch pools' where they can handle live creatures like starfish and crabs. Fun features include those which gives the visitor a 'fish eye view' of a waterfall and a deep submarine vehicle.

The local tourist offices provide detailed maps and guides to help tourists to get around. Information is available on the many types of accommodation from hotels to hostels. Walking trails have been designed to help you make your way around this compact city.

There is an excellent transport network linking the city to tourist attractions in the locality. The train station is located in the heart of the city, Eyre Square.

Key points

- **Tourism** is hugely important to the Irish economy.
- **Tourism** is particularly important in areas where there are few opportunities for work.
- Cities offer many attractions and facilities for **tourists**.

Tourism in Ireland: Tourist Regions

Key words

National park
Protected wilderness

Activity-based holiday
Recreational

Donegal's dramatic coastline

Glenveagh National Park

Glenveagh National Park is an excellent example of where tourist facilities are provided and carefully managed. National parks have been created to serve both tourists and locals. Experts are employed to plan and manage this important resource.

This national park annually attracts over 100,000 visitors. Traffic is carefully managed. A car park is provided near the entrance to the park. A shuttle bus transports tourists around the park. Restaurant facilities and guided tours are provided in the castle at Glenveagh.

North-west Donegal: an area of natural beauty

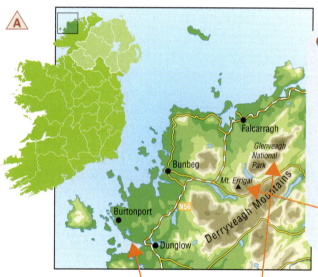

● The dramatic peak of Errigal Mountain (725 metres high), snow-capped in winter.

Donegal is an area of great natural beauty. Its attractions include:

● The U-shaped glaciated valley of Glenveagh. This **national park** has a magnificent castle and vast acreage of **protected wilderness**. It is home to Europe's largest remaining wild red deer herd.

● Beaches backed by dramatic cliffs, caves, arches and stacks provide spectacular scenery for the tourist.

Glenveagh National Park

Beaches and recreational attractions

Wexford has a long coastline and many beaches. This area is described as the 'Sunny South-East' and with good reason. It is the sunniest corner of Ireland. Read the postcard to discover some of the many attractions of the coastal resort.

Hi Jenny,

We arrived safely. It took us a while to find the mobile home. We spent the next day on the beach which is only a stone's throw away. Decided to eat out last night in the local hotel. Enjoyed beautiful food and great entertainment there. Yesterday I treated myself to a relaxing day in the recreation centre in the hotel. Sauna, massage - the works you might say. And yes! It's true about holiday romances. Met the man of my dreams at the local disco. Planning to go wind-sailing tomorrow with him. May never go back to the big smoke.

Love Tess

Jenny Murphy

32 Main Street

Newbridge

Co. Kildare

The south-west: activity-based holidays

One of the most popular areas for an **activity-based holiday** is the Ring of Kerry. Recreational and sporting facilities include the world famous golf courses at Ballybunion and Waterville.

Hill walkers and hikers regularly climb Ireland's highest mountain, Carrauntohill and other peaks in the Macgillycuddy Reeks range.

South-west activity based holidays

There is a plentiful supply of tourist facilities. There are adventure centres offering activities ranging from abseiling to windsurfing. Maps and walks are widely available along with plenty of accommodation and entertainment.

Government money has been made available to provide other facilities for tourists, for example, the Water Park at Tralee.

Activity-based holidays are largely the result of local enterprise. By providing services and facilities for tourists, the most can be made of the attractions supplied by nature.

Key points

- Ireland offers different attractions, including beaches and **recreational** facilities.
- A range of facilities have been put in place to serve the needs of the tourist.

177

The Role of Climate on Tourism: Spain's Tourist Industry

Key words

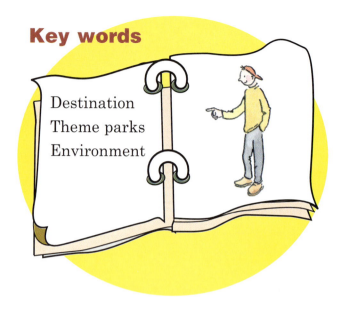

Destination
Theme parks
Environment

Spain is the second most popular tourist **destination** in the world. France is the most popular. About 53.1 million tourists visited Spain in 2003. The south of Spain is the most popular region. Visitors come to enjoy its warm, sunny climate and golden beaches. The Mediterranean coastline of Spain and its islands is the most popular destination for Irish tourists. It accounts for 60% of all Irish foreign holidays.

Warm, sunny climate

The south of Spain has a warm, sunny climate.

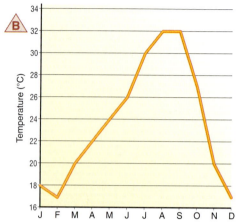

Average monthly temperatures in Mediterranean Spain

Southern Spain enjoys over 320 days of sunshine per year, extremely low rainfall and mild temperatures. Even in December the temperature rarely drops below 17°C. The information in chart B is a guide to the average temperatures you may expect while on holiday in the Mediterranean areas of Spain.

Palma de Mallorca (Majorca)

One popular resort is Palma de Mallorca (Majorca). It is located in the Balearic Islands off the east coast of Spain in the Mediterranean Sea.

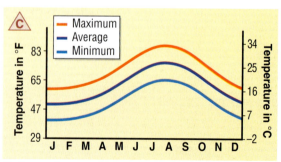

Palma de Mallorca climate graph

The chart shows that Palma de Mallorca's temperatures rarely fall below 8°C in winters or 20°C in summer.

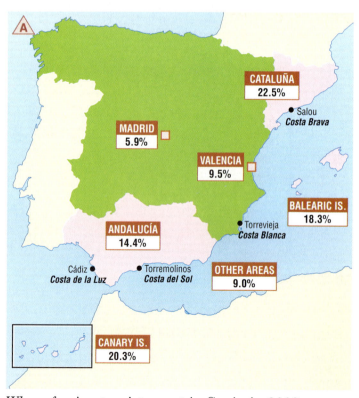

Where foreign tourists went in Spain in 2003

The precipitation chart shows that rainfall is low especially during the summer months. This means clear cloudless skies, perfect tanning weather!

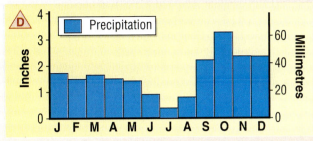

Rainfall in Palma de Mallorca

This attractive climate means that open-air **theme parks** can be enjoyed without the threat of rain in the summer months.

E Visitors to Spain in 2003					
Domestic	UK	Germany	France	Rest of Europe	Rest of World
46%	17%	10%	8%	16%	3%

Source: Spanish National Official Statistics

The table shows that almost half of all tourists in Spain are the Spanish themselves. In summer, they leave the very hot areas inland and go to the coast. The single most popular destination is the Mediterranean coastline. Look at map A to see the most popular resorts.

Impact of tourism on the Spanish environment

In Spain, mass tourism has brought overcrowded towns, traffic congestion and noise and water pollution. High-rise, high-density apartments have been built along the Mediterranean coastline changing the look of villages. The high density of people in the summer months adds to the problems of untreated water sewage. This has had harmful effects on the **environment**.

Impact of tourism on society

Tourism may have an unwelcome effect on society. Traditions may be replaced to please tourists who are interested in a 'home-from-home' type holiday. Family meals or siesta periods may be altered in keeping with the ways of visitors. Petty crime levels may increase as greater numbers of wealthy tourists visit the area.

Key points

- Tourists mainly go to Spain because of its sunny climate.
- Facilities such as **theme parks** take advantage of the warm sunny weather.
- Tourism can have an unwelcome impact on society and the **environment**.

Tourism, Transport and Communication Links

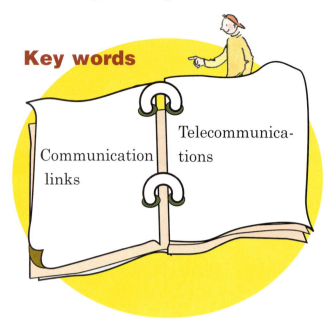

Key words

Communication links

Telecommunications

To provide for tourists, money is spent improving transport and other **communication links**. Let's look at Switzerland to show the link between tourism and transport.

Case Study: Switzerland

Switzerland has one of the oldest tourist industries in the world. In the past, wealthy people, suffering from illnesses such as TB, travelled to benefit from the clear mountain air of Switzerland. The Swiss tapped into this wealthy market. But they took time to develop the tourist industry on a planned basis. Part of this planning involved the building of excellent transport and communication links. This was a challenge in a country largely made up of mountains.

Switzerland – its landscape

Switzerland has a hilly landscape. About 100 peaks are close to or higher than 4,000 metres above sea level. It has many rivers and lakes, the largest of which is Lake Geneva. The rivers of Switzerland include the Rhine and Rhone. The Alps make up some 60% of the surface of the country.

Rail and routeways in Switzerland

Transport in Switzerland

Switzerland has over 6,000 km of railway lines including numerous railway tunnels and three road tunnels through the Alps. Fast trains depart hourly from the airport railway stations of Zurich and Geneva, for up to 16 hours a day.

Hourly rail departures also operate between all major Swiss cities. An extensive network of nearly 500 mountain railways takes you up to the snow-capped summits of the Swiss Alps.

Rail links criss-cross Switzerland

By road

Switzerland has over 70,000 km of road networks. A network of well-marked motorways spans the country. The bright yellow Postbus buses take visitors to the remotest regions of the country.

All aboard!

By waterway

Nearly 170 vessels, many carrying tourists, journey along more than 20 waterways (rivers and lakes) throughout the country.

Barging on the river

Most tourists use the Swiss Travel System. The system is as reliable and punctual as a Swiss watch. The timetables are perfectly linked. This leads to trouble-free changes from one means of transport to another.

Telecommunication links

Telephone lines and mobile phone coverage has been developed to a high standard in Switzerland. **Telecommunication** links are very important in supporting tourism.

Two-way traffic

Tourism encouraged the development of an excellent transport and communication network in Switzerland. In turn, this excellent network encourages even more tourists to visit.

Impact of tourism on the Swiss environment

As Switzerland is a landlocked country many tourists come by car. They drive up through the mountains and their cars belch out fumes as they move. This damages the trees and increases the amount of acid rain.

Building a wide network of ski lifts also damages the wildlife. Then, with the many visitors trampling over the same area, the soil is loosened. This causes it to move. Landslides or avalanches can occur.

Ski-lifts damage wildlife

Key points

- Switzerland's excellent transport network was built in response to the demands of tourism.
- Good **communication links** benefit tourism.
- Tourism can have an unwelcome impact on the environment.

REVISION EXERCISES

Write the answers in your copybook.

1   Read these statements about tourism. Which are true?
   ● Tourism is a tertiary activity
   ● The number of tourists in the world is decreasing each year
   ● Glenveagh National Park is in Co. Wexford
   ● Tourism brings jobs and money to an area

2   What is the main reason for the growth of mass tourism in the Mediterranean area?
   ● Sunshine
   ● Winter sports
   ● Cities
   ● Natural parks

3   *Tertiary activities now employ the highest percentage of the workforce in Ireland.*
   (a) Explain what 'tertiary activities' are.
   (b) Name **three** tertiary activities.

4   List **two** benefits that tourism brings to an area. Explain **one** way tourism has benefited a named region in Ireland.

5   Tourism also brings problems to an area. Describe **two** problems often experienced in areas of mass (great) tourism.

6   Tourism can be a major reason why some areas are conserved. Some areas of conservation are called national parks.
   (a) What are the **two** aims of national parks?
   (b) Explain **one** way in which tourism helps to conserve these areas.

7   The following table shows where most tourists to Ireland come from. Imagine that you are trying to increase the numbers visiting. Which place would you target? Give **one** reason for your choice.

Britain	Mainland Europe	USA	Other Areas
57%	24%	13%	6%

182

8   Examine the percentages showing where most tourists go once they arrive in Ireland.

(a) Give **two** reasons why 27% visit the Dublin area.

(b) Suggest **one** reason why only 8% visit the North-west region.

Dublin	Midlands/East	South-east	South-west	Shannon	West	North-west
27%	9.8%	12%	17.2%	11%	15%	8%

9   Look at the bar chart showing the number of tourists who visited an historic building in a year. The largest number visited the building in:

● February   ● April

● August   ● December

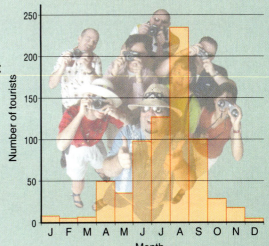

10   *Climate attracts Irish people to tourist areas in many European countries for their holidays.*

(a) Name **one** such tourist area you have studied.

(b) Explain why the climate attracts tourists.

11   Ireland's main tourist resources include:

● Cities

● Coastal resorts

● Sport and recreational areas

● Areas of natural beauty

Choose **two** of the above resources and for each resource you have chosen explain why it is important in attracting tourists to Ireland.

12   Name a Mediterranean area you have studied that is important for tourism.

(a) Give **two** reasons why it has become important.

(b) State **one** problem caused by large scale tourism in such areas.

13   *Switzerland's transport network is linked.*

(a) Explain what that means.

(b) Detail **one** way that this helps the tourist.

14   The following chart shows the precipitation (mainly rainfall) levels at Palma in Mallorca.

(a) What month receives the most rainfall?

(b) What month receives the least rainfall?

(c) What season has the highest amount of rainfall?

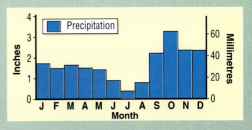

183

20 Industrialised Society and the Environment

Key words

Fossil fuels
Acid rain
Coniferous
Acid soils
Alternative energy

Let's look at look at some serious side effects of development.

Acid rain

When **fossil fuels** (such as coal, oil and gas) are burned, gases are sent up into the atmosphere. These gases are sulphur dioxide (SO_2) and nitrogen oxides (NOx). These gases mix with rainwater, oxygen, and other chemicals to make **acid rain**. Acid rain damages the important natural resources of trees, lakes, rivers, and soils.

Damage to trees

Seven million hectares of European forest are dead or dying because of acid rain. The figures below show you the percentage of the trees that have been damaged by acid rain.

A	Country	% of Forest Dead or Dying
	Germany	54
	Switzerland	50
	Netherlands	50
	Poland	27
	Austria	25

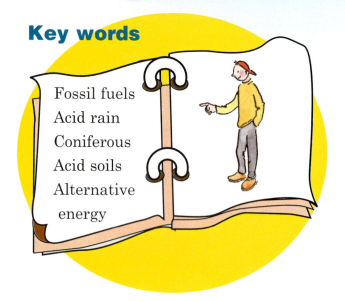

80% of the lakes in southern Norway have acid rain pollution of which 50% comes from Great Britain

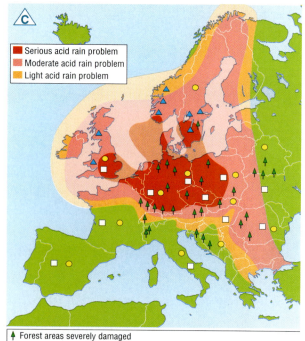

C

■ Serious acid rain problem
■ Moderate acid rain problem
■ Light acid rain problem

🌲 Forest areas severely damaged
▲ Lakes and streams damaged by acid rain
● Country in which more than 1 million tonnes of sulphur dioxide falls in a year
□ Country that releases at least 1 million tonnes of sulphur dioxide annually

The effects of acid rain in Europe

When acid rain seeps into the ground, toxic (deadly) minerals build up in the soil so that trees do not grow well. They become stressed and are now more likely to suffer damage from insects, fungi, frost, wind and drought.

Coniferous trees, pine and spruce, showing damage caused by acid rain

Farming

Acid soils are infertile and are unable to support a healthy crop. The soil in parts of Scandinavia is now 10 times more acidic than 50 years ago.

The built environment

Acid rain attacks metal and stonework. The major threats are to older historic buildings. Many monuments like the High Cross of Cong (Mayo) and the Acropolis (Greece) have been damaged by acid rain.

The Acropolis in Athens, Greece, shows signs of damage from acid rain

Who is to blame?

Developed countries are to blame for most of the damage caused by acid rain. They burn vast amounts of **fossil fuels** to produce energy for industry, homes and transport. The average American uses twice as much energy as a European and 1,000 times as much as someone from a developing country such as Nepal.

It is difficult to find a solution as acid making gases that are produced in one country may be carried by the wind into another country.

Some solutions

Devices called scrubbers should be fitted to chimneys in electricity stations where fossil fuels are burned. They can remove the SO_2 from the gases burned there.

Use **alternative** (other) **energy** sources such as hydropower, wind energy, geothermal energy, and solar energy. These are 'cleaner' fuel sources.

Conservation saves energy by using less of it.

Key points

- When **fossil fuels** are burned they release harmful gases into the air.
- These harmful gases mix with rain and fall to the earth as **acid rain**.
- **Acid rain** causes serious damage to soils, trees, rivers and the built environment.
- The solution to the problem of acid rain includes using **alternative energy** sources and conservation.

The Greenhouse Effect

Key words

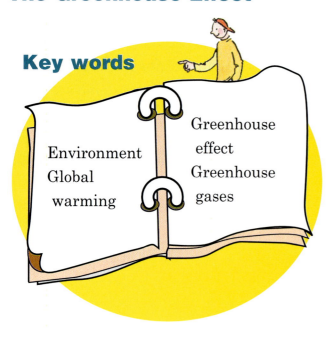

Environment
Global warming

Greenhouse effect
Greenhouse gases

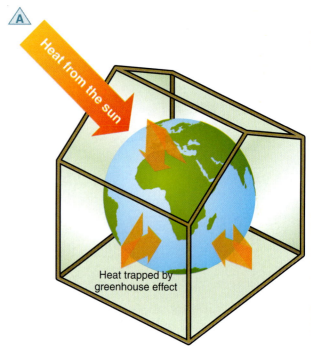

With more of these gases in the atmosphere, the air will be warmer

One way in which we have damaged the **environment** is through **global warming**. This is also called the **greenhouse effect**.

Heating the earth's atmosphere

A blanket of air, the atmosphere, surrounds the earth. It is a mixture of gases, water vapour and dust. The rays of the sun strike the atmosphere as they make their way to the surface of the earth. About 30% of the energy is reflected straight back into space, but the rest passes down through the atmosphere and heats the ground. The ground then sends the heat back into the air like a giant radiator.

Greenhouse gases

More heat is being trapped in the atmosphere so global air temperatures are rising. This is caused by the increase in levels of carbon dioxide and other gases in the air. These gases are good at trapping heat, which is why they are called greenhouse gases. Think of a greenhouse (or glasshouse) and how heat is trapped inside.

The level of **greenhouse gases** in the atmosphere is rising because of human activity. Let's look briefly at these gases.

1. *Carbon dioxide*: The burning of fossil fuels adds CO_2 to the atmosphere. By 2020, if present trends continue, the amount of CO_2 in the atmosphere will be double its natural level.

2. *Methane*: The amount of methane in the atmosphere is increased when forests are cut down and replaced by grassland. Termites living in, and cattle grazing on these grasslands add more methane to the air than the wild animals of the original forests.

3. *CFCs*: They are used in aerosols, refrigerators and foam packaging. CFCs are very effective greenhouse gases.

Therefore human activities are making the air warmer.

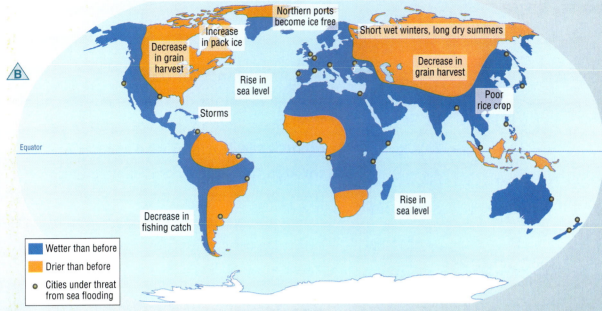

What might happen by the year 2050 if global warming continues at current rates

Problems caused by global warming

Rising sea levels

As global temperatures rise, the polar icecaps melt, adding water to the oceans. As the level of sea rises, low-lying land is flooded. Scientists believe that the sea level has risen by about 10–25 cm over the last 100 years. The rise in sea levels is very worrying as most of the world's population lives close to the coast on low-lying land.

Severe storms

There is also the possibility that unusual weather events, such as severe storms, may become more frequent. The increase in heavy downpours and flooding in Ireland in recent times has been blamed on global warming.

Increase in tropical diseases

With higher temperatures worldwide, tropical diseases, such as malaria, could become more widespread.

Solutions to global warming

Key points

- The level of **greenhouse gases** in the atmosphere is rising because of human activity.
- More heat is being trapped in the atmosphere so global air temperatures are rising.
- Global temperature rises cause the level of the sea to rise due to melting of polar icecaps.

The Ozone Layer

Key words

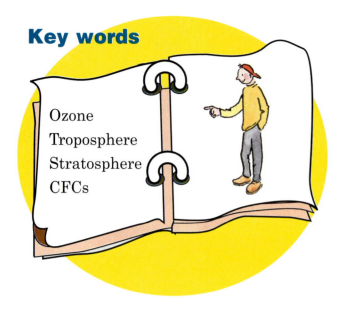

Ozone
Troposphere
Stratosphere
CFCs

A hole has opened up in the **ozone** layer of the earth. The result is that more of the sun's rays can enter the atmosphere. This has two effects:

- The air heats up more.
- Harmful rays causing cancer can pass through the air.

What is the ozone layer?

The *earth's atmosphere* is made up of several layers. The **troposphere** is the layer of air closest to the earth. This is the 'weather' layer where there is wind, cloud and rain. This lowest layer rises from the earth's surface to about 10 km. (Mt Everest, the highest mountain on the planet, is about 9 km high.)

Above the troposphere is the **stratosphere**, which continues from 10 km to about 50 km. This is where you find the ozone layer. Ozone is a very important gas. Picture it as a layer of 'sun-screen' which blocks out some heat and harmful ultraviolet rays.

The importance of ozone

Human activities destroy ozone. When this happens a 'hole' or gap opens up in the ozone layer. Harmful rays can pass through this hole. Ultraviolet radiation from the sun harms most living things. We can be unaware of skin cancers and eye cataracts caused by these rays until it is too late.

CFCs destroy ozone

We know that CFCs found in refrigerators, solvents, industrial chemicals and fire extinguishers damage the ozone layer. Vehicles emit more CFCs than all other culprits together. They leak CFCs into the **troposphere**. Over time, winds drive the CFCs into the **stratosphere**. The damage to ozone is worse at the North and South Poles because of weather conditions there.

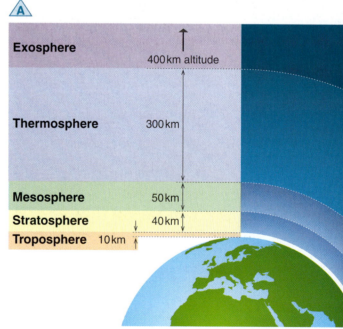

A

The earth's atmosphere is composed of several distinct layers. The ozone layer is in the stratosphere

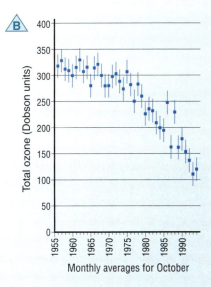

Springtime ozone depletion (loss) in Antarctica 1955–94

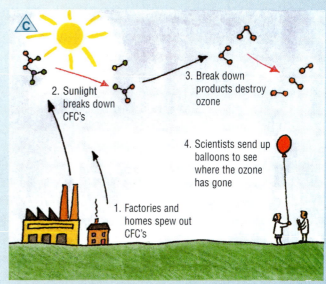

CFCs destroy ozone

The hole over the Antarctic

There is a big hole in the ozone layer over Antarctica at the South Pole. The conditions of extreme cold and darkness in winter followed by bright conditions in spring and summer cause chemical reactions that destroy ozone.

The graph shows that the amount of ozone over Antarctica in spring has fallen dramatically between the years 1955 and 1994. This is very worrying because of the damage it does to our health.

It is also worth noting that some of the gases that destroy ozone are also greenhouse gases that you learnt about in the last section.

Tackling the problem

We know that using aerosol sprays produces CFCs. In the 1970s there was a ban on the use of CFCs in aerosols in several countries, including America.

CFCs are produced when we burn petrol in cars. Cutting down on the use of cars would lower CFC levels. Looking for alternative sources of energy to run cars is another option.

If no more damage is caused it is hoped that the ozone layer will be able to heal itself over a period of about 100 years.

Key points

- The **ozone** layer acts as a natural sunscreen. When it is damaged, holes appear and harmful rays get through.
- Human activities such as using **CFCs** have damaged the **ozone** layer.
- The banning of **CFCs** and use of other fuels will help the **ozone** layer to heal.

REVISION EXERCISES

Write the answers in your copybook.

1   Trees like the ones in the photograph
    are killed by:
    ● Acid rain
    ● The ozone layer
    ● Abrasion
    ● Solar energy

2   Acid rain is formed when:
    ● Trees are planted in rainforests
    ● Fossil fuels like coal are burned
    ● Glaciers melt
    ● Limestone is weathered

3   The rapid warming of the earth's atmosphere due to the burning of
    fossil fuels is known as:
    ● Acid rain
    ● The greenhouse effect
    ● Pollution
    ● Erosion

4   Draw a diagram showing the main causes of acid rain.

5   Write a short account of the damage done by acid rain to:
    ● Forests
    ● Farms
    ● Buildings

6   Describe **two** of the ways in which greenhouse gases are built up in
    the atmosphere. Illustrate your answer using a carefully labelled
    diagram.

7   Explain what the term 'greenhouse effect' means.

8   Write the correct answer for **each** of the following
sentences in your copybook.

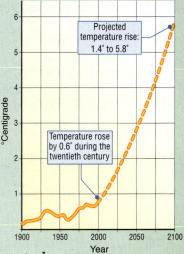

(a) During the twentieth century temperatures rose by
1.4°C / 0.6°C.

(b) By the year 2100 temperatures are expected to rise by
between **0.6°C to 1.4°C /1.4°C to 5.8°C**.

(c) The main cause of global warming is the increase in
CO₂ emissions /acid rain.

9   Copy the above diagram (question 8) into your copybook.

(a) By how many degrees Centigrade is the temperature expected
to rise in 2100?

(b) Describe **one** change that this temperature rise could bring to
an area.

10 Explain how **two** of the following might help to ease the problem of
global warming.

● Energy conservation
● Use of renewable energy sources
● Increasing the size of areas planted with trees

11 Explain **one** benefit of global warming.

12 The following diagram shows how the
ozone layer is damaged. Copy it into your
copybook.

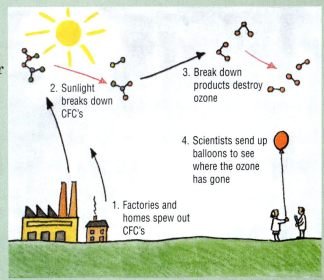

(a) Explain where CFCs come from.

(b) Describe the way in which they
damage the ozone layer.

(c) What can be done to lessen the
damage in the future?

13 Without the ozone layer we would suffer. Explain **two** of the ways
in which the lower amount of ozone is harming human life.

21 A World of Differences

Key words

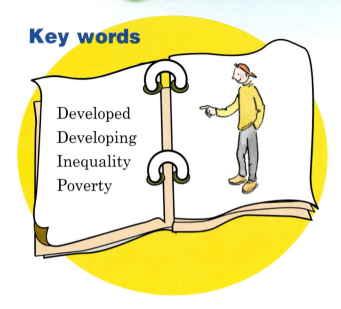

Developed
Developing
Inequality
Poverty

The resources or riches of the world are not evenly shared out. Some people are very rich, other people are very poor. You find differences in the share of the world's wealth *between* different parts of the world and *within* countries themselves.

The world's wealth – who has it?

The wealthiest countries in the world are the **developed** countries. They include the countries of Europe, North America and Japan. Because most of them are in the northern part of the world, developed countries are described as countries of the 'north'. Although a country like Australia is in the southern half of the world, it is still part of the 'north' because it is developed.

Of course **inequality** exists *within* countries. Not everyone in America for example is well off. But it is true that even poor people in rich countries are not as poor as poor people in poor countries. The poorest people in the world are said to be part of the **developing** world. These countries are sometimes referred to as the 'south' because most of them are in the southern half of the world. Life in many of these countries is improving but only very slowly. These countries can be described as *slowly developing*.

Some poor countries are now beginning to move out of their **poverty**. Their economies are taking off. Although still poor, they are described as *quickly developing* countries.

B	Developed	Slowly Developing	Quickly Developing
	Ireland	Mali	Brazil
	USA	Afghanistan	China

Developed
Quickly developing
Slowly developing
North-South divide

The dividing line between 'developed' and 'developing' countries

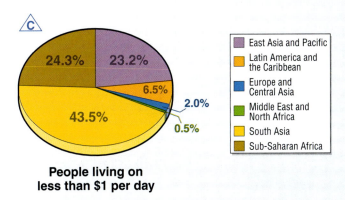

People living on less than $1 per day

Legend:
- East Asia and Pacific
- Latin America and the Caribbean
- Europe and Central Asia
- Middle East and North Africa
- South Asia
- Sub-Saharan Africa

The percentage of people in different parts of the world who have to survive on less than a dollar per day. You can see that almost half of the world's poor live in South Asia

Economic activities

From the example of China you learnt that as a country becomes more developed, more and more of its people move from employment in primary activities, such as farming, to employment in secondary activities (factory work). As a country develops further, more people move into jobs in the tertiary or service sector.

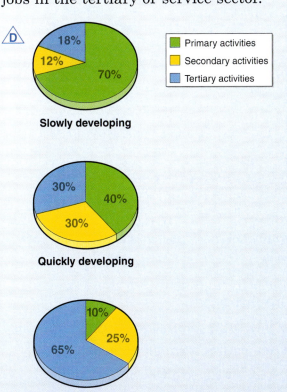

Legend:
- Primary activities
- Secondary activities
- Tertiary activities

Slowly developing

Quickly developing

Developed

People employed (%) in primary, secondary and tertiary activities in slowly developing, quickly developing and developed countries

Widening gap between rich and poor

In the past decade poverty has been rising in Africa, in South Asia, and most dramatically in Eastern Europe.

While this is happening, the rich in the rich part of the world are getting richer. The richest 20% in the world have become richer, while the poorest 50% have become poorer.

It's not fair!

According to a recent report some 400 super-rich Americans had an average income of nearly $174 million each, or a combined income of $69 billion, in 2000. Incredibly, that's more than the combined incomes of the 166 million people living in Nigeria, Senegal, Uganda and Botswana. America's richest individuals could actually change the course of Africa's history.

Key points

- The resources of the world are not evenly shared.
- Some countries, the **developed** ones, are very well off compared to the slowly **developing** (poorest) countries.
- The economies of some **developing** countries, including China, are now quickly **developing**.
- The gap is widening between the rich and poor countries of the world.

The Widening Gap

Key words

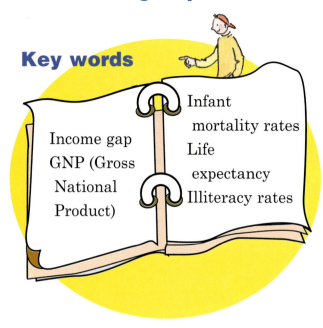

Income gap
GNP (Gross National Product)
Infant mortality rates
Life expectancy
Illiteracy rates

A

That makes an average of 20 euro each.

THE AVERAGE POCKET MONEY IN THIS CLASS IS 20 EURO

But I get 35 euro

I only get 5 euro

Let's look a little more closely at the differences or gaps between rich and poor countries. It's a gap that's getting wider. This **gap** is seen in the areas of **income**, health and education.

The income gap

It is possible to work out the average income of people in a country. The total value of all the goods and services in a country is added up and divided by the number in the population. This gives us an average income figure. It is sometimes called the **Gross National Product (GNP)**. It is usually given in US dollars ($).

This figure allows us to compare countries. Income per person in the world's 20 poorest countries has barely changed in the last 40 years. It was $212 in 1960–62 and has risen to $267 in 2000–02. Income in the richest 20 nations has tripled, from $11,417 to $32,339 over the past 40 years.

B

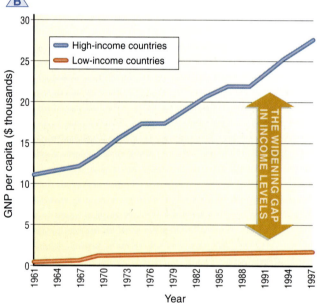

The gap between the incomes of rich and poor is widening.

The health gap

There is a huge gap between rich and poor countries across a range of health measures. In the United States, public spending is about $2,000 for every person every year. In Africa, public health spending is around $10 per person per year.

Infant mortality rates

The 'health gap' between the developed and developing world is seen very clearly when we look at the **infant mortality rates** (the number of children who do not survive past their first birthday). Look at chart 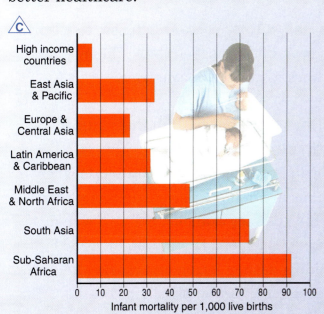 C. Notice the difference in the rates for the high-income countries and the low-income countries of Asia and Africa. The number of children who die around the time of childbirth is 20 times higher in sub-Saharan Africa and South Asia than in the rich industrial countries. Most of these deaths in low-income or developing countries could be prevented if there was better healthcare.

Life expectancy

Most of us in the developed world can expect to live until our late seventies. In Africa **life expectancy** is less than 50 years. It is less than 40 years in some of the AIDS affected countries of Africa.

The education gap

The gap in education standards between the developed and developing world is another sign that the world's wealth is not evenly spread. Look at chart D and notice the differences.

In many poor countries, only half of the children of secondary school age are at school. In many developing countries there are **illiteracy rates** of nearly one-third.

C

High income countries
East Asia & Pacific
Europe & Central Asia
Latin America & Caribbean
Middle East & North Africa
South Asia
Sub-Saharan Africa

0 10 20 30 40 50 60 70 80 90 100
Infant mortality per 1,000 live births

The health gap

D

Sub-Saharan Africa
South Asia
East Asia
South East Asia & Pacific
Latin America & Caribbean
Eastern Europe & CIS
Industrial countries

0 10 20 30 40 50 60 70 80 90 100
% of children in secondary education

Education inequality

E

Country	GNP per person 2003 $	People per doctor 2003	Literacy levels 2003 %
Ireland (developed)	26,960	632	99
Brazil (quickly developing)	2,710	844	86.4
Mali (slowly developing)	290	18376	19

Key points

● There is a widening **gap** in **income**, health and education between developed and developing countries.

Unfair Trade!

Key words

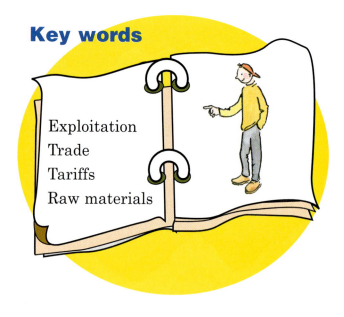

Exploitation
Trade
Tariffs
Raw materials

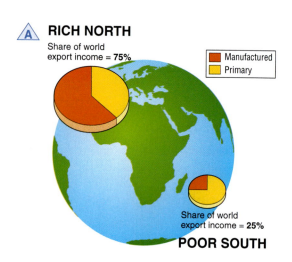

A RICH NORTH
Share of world
export income = **75%**

Manufactured
Primary

Share of world
export income = **25%**
POOR SOUTH

Many poor countries are poor because they have been or are being exploited by richer and more powerful countries. When we speak of **exploitation**, it means that one country takes advantage of, or oppresses, another.

This oppression is clear when you look at the issue of **trade**. The exchange of goods is called **trade**. Goods bought from other countries are called imports. Goods sold to other countries are called exports. Sometimes **tariffs** or taxes are added to the price of imports.

Unfair trade

The scales are tipped in favour of the rich countries when it comes to trade. They set the terms or conditions of trade. These terms favour the rich.

Tax on imports

Some developed countries put taxes on imports from developing countries. This pushes up the price of the goods. It makes it difficult for poorer countries to sell their goods in the market place. These terms work against poor countries.

Raw materials, not manufactured goods

Value is added to **raw materials** when they are manufactured into something else. Take fish for example. Sell it raw and you might earn €1. Dress it up in garlic butter, add a dash of lemon and charge the customer €3. You have added value to the product in the factory.

For this reason developed (rich) countries prefer to buy goods in a raw state. They want to add value to raw materials in factories and make more money. So they are willing to buy raw materials such as cocoa at a low cost with a plan to turn it into a valuable product such as chocolate. They are not willing to buy manufactured goods from developing (poorer) countries. This forces the developing countries to rely on the sale of less valuable raw materials.

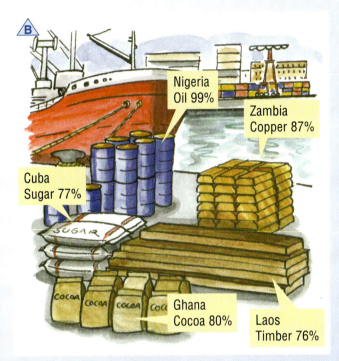

Single export countries

Trade in money

To fund development, many developing countries have borrowed large sums of money from international banks. They have to pay back these loans with the added interest. This trade in money has kept many poor countries in poverty. It has been described as a 'debt trap'. Until these debts are written off, the poor countries will remain poor.

Many now believe that a change in the terms of trade is one of the key ways to lessen poverty in these poorer countries. Barriers to trade include taxes on imports.

In many cases the poorer countries do not get a fair price for the raw materials they sell. The case study on coffee later in this section will show this.

Depending on one cash crop

Many poorer countries depend on the sale of one cash crop. If this crop fails or the price crashes on world markets they struggle to survive. The price paid for this product is often decided at meetings of powerful business people. They will take it at the lowest price possible, which is not in the interest of poorer countries.

The burden of debt

Increasing their share of world trade

Some rich countries have set a limit on the amount or value of goods that can be traded from poor countries. If the richer countries allowed a mere 1% increase in share of world exports by Africa, East Asia, South Asia, and Latin America, it would result in an income rise that could lift 128 million people out of poverty. In Africa alone, this would make $70 billion – about five times what the continent receives in aid.

Key points

- Terms of **trade** arrangements appear to benefit the rich countries of the world.
- Because of unfair trade the poorer countries find it really difficult to develop.

Ireland in the Past – Exploitation

Key words

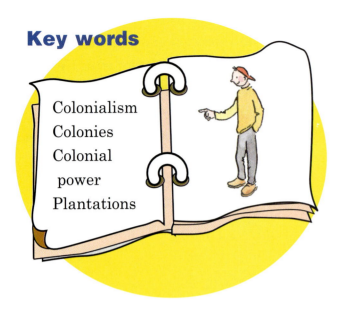

Colonialism
Colonies
Colonial
power
Plantations

In common with many poor countries in the world, Ireland has a history of **colonialism**. Colonialism is a system that exploits (takes advantage of) weaker countries and keeps them poor.

A history of colonialism

Many of the developing countries were ruled by European powers in the past. They were known as colonies and their rulers known as **colonial powers**. The attitude of the colonial powers was that the colony was there to make them richer. They took over the land and set up large farms called **plantations**. They grew crops for export. These crops, known as cash crops, were the raw materials they needed to manufacture the goods that would make the colonial power rich. Local people worked as labourers on the large farms. They were paid a very low wage. As a result of colonialism these poorer countries found it difficult to make progress.

Ireland's colonial past

This is also Ireland's story. In the past Ireland was a colony of England. This means it was controlled by England. Colonisation kept Ireland poor.

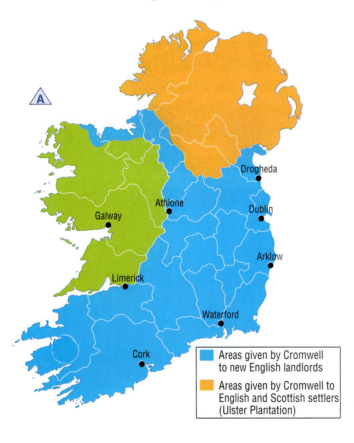

Areas of Ireland that were colonised by the English

Protection of English goods

The colonisation of Ireland – similarities with developing countries today

Industrial England needed food to feed its people. It also needed farm products to be used as raw materials that could be processed into manufactured goods. Ireland, being in her 'back yard' could supply these products. England did not want competition from goods made in Ireland. As a result, the English did not invest in factories here. England wanted to make sure that people could only buy manufactured goods from their factories in England.

Many developing countries supply cheap raw materials to developed countries. For a long time industry was not encouraged in developing countries. Like Ireland, this slowed up their development.

It was difficult to sell Irish made goods in England because an import tax made them expensive. This kept Ireland poor.

Developing countries find it hard to break into markets because of taxes put on manufactured goods.

As a colony Ireland had to buy from England even if their prices were not the cheapest. This benefited them but it kept Ireland poor.

Many poor countries are forced to sell at the price decided by richer countries.

During the plantations era the planters set up large estates. The local Irish men and women worked as tenants or labourers on these farms. Many of the crops grown were exported as cash crops. The profits went back to England (where the landlord may have lived) or the money was used to plan and build towns.

Many poor people in developing countries work for very low wages on large plantations that produce cash crops for export.

For those Irish who owned a farm, life was hard. The plots were small and were often in areas of bog or infertile soil. They depended on potatoes as their staple food. When that failed, the Great Famine happened. Even while the people starved, food crops were exported to England.

This is the situation in many poor countries where people go hungry.

Key points

- Like many developing countries, Ireland has a history of **colonialism**.
- The system of **colonialism** – where one country controls another – held Ireland back.
- Developing countries are poor because of their history of **colonialism**.

The Coffee Trade

Key words

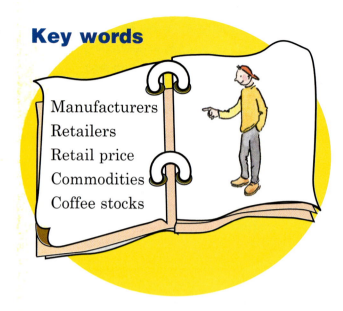

Manufacturers
Retailers
Retail price
Commodities
Coffee stocks

Let's look at the coffee trade. It shows very clearly how many rich countries exploited and still exploit poorer countries.

Key words explained!

Knowing the meaning of the following words will help in your study of the coffee trade.

Manufacturers: The companies that process the coffee (roast and pack it).

Retailers: The companies that sell the coffee.

Retail price: The price for which something is sold in the shops.

Commodities: Things that are bought and sold on the world market.

Coffee, please!

When you drink your next cup of coffee, think about how much it cost. Then ask yourself:

How much of that price went to the growers, the people who planted the coffee, picked the coffee beans, cleaned them, and dried them? How much went to the exporters? How much went to the companies who shipped the coffee beans and roasted them? How much went to the retailers who sold the coffee in their shops?

Coffee – a valuable commodity

Coffee is the second most valuable commodity after crude oil. It is the most valuable farm commodity in world trade. Coffee is a multi-million dollar industry. Despite the popularity of coffee in the countries of the north, many of the coffee growers in the south live in extreme poverty. These farmers sell their crops to multinational companies in its raw stage when profit is lowest. The multinational companies who control the shipment, roasting and sale of these commodities, earn the most from the sale of coffee.

The price of coffee can change very quickly – this affects price. It depends on:

● **Coffee stocks** worldwide. A fall in stocks means a high price.

● Political conditions – a war may affect trade.

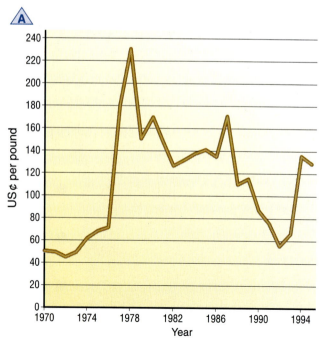

Changes in the price of coffee over time

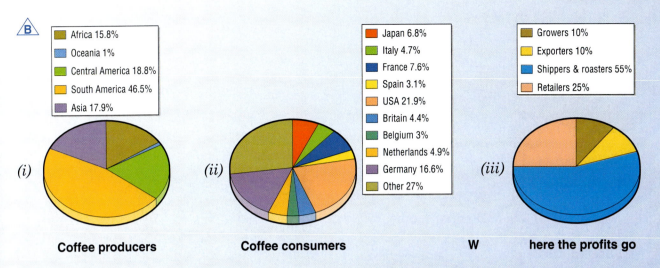

Africa 15.8%	Japan 6.8%	Growers 10%
Oceania 1%	Italy 4.7%	Exporters 10%
Central America 18.8%	France 7.6%	Shippers & roasters 55%
South America 46.5%	Spain 3.1%	Retailers 25%
Asia 17.9%	USA 21.9%	
	Britain 4.4%	
	Belgium 3%	
	Netherlands 4.9%	
	Germany 16.6%	
	Other 27%	

(i) Coffee producers   *(ii)* Coffee consumers   *(iii)* W here the profits go

(i) Coffee producing countries; (ii) Coffee consuming countries; (iii) How the profits of coffee production are shared

Chart **A** shows how world coffee prices suddenly rose as a result of serious damage to the Brazilian coffee crops (20% of the world's coffee) in 1975 (frost), 1984 (drought), and 1994 (frost).

Who buys the coffee?

When the price of raw coffee rises on world markets, the coffee drinkers pay more for their cup of coffee. But the producers do not earn more. Most of the profit goes to the four leading manufacturers who control the world coffee market. These are Nestlé, Procter & Gamble, Kraft and Sara Lee.

Unfair trade prevents development

In some poor countries the sale of coffee makes up a huge part of their total income. Burundi in Africa, for example, earns 80% of its foreign earnings from the sale of coffee alone. When the world price of coffee falls, they earn less. Meanwhile the price of their imported manufactured goods continues to rise. This means they can buy less: their money doesn't go as far.

C

I have to trade this much to be able to import a lot less. It seems so unfair

Imports

Exports

Unfair trade – the exchange is unfair

Key points

- In many cases, coffee producers do not get a fair price for their crop.
- Most of the profit from coffee growing goes to business people in the developed world.
- Unfair prices for the coffee crop keeps poor people poor.

What Kind of Aid?

Key words

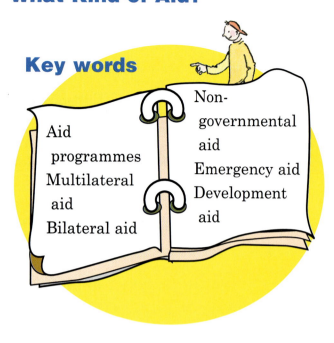

Aid programmes
Multilateral aid
Bilateral aid
Non-governmental aid
Emergency aid
Development aid

It is important to remember that the wealth of the world is there for all to enjoy. It is unfair that some people have so much and others have so little. Aid is about working towards a more equal sharing out of wealth.

An unfair share

Many aid workers can see the huge difference that a little help can make. A donation of money can mean a new well is opened up and water is brought to the fields. Crops can be grown and a village can prosper.

Sharing is fairer

Let's examine the many ways in which resources are shared. With the help of various **aid programmes**, we are creating a more equal world.

Sources of aid

- **Multilateral aid:** This is where governments give aid to a central international organisation like the United Nations (UN). The UN decides where the money is best spent.
- **Bilateral aid:** This is where *one* government gives directly to another government for development projects.
- **Non-governmental aid:** This is aid given by voluntary groups such as Self Help or Concern who raise money and undertake projects in the poorer areas of the world.

Types of aid

- **Emergency aid:** This is short-term aid. It is usually given in times of crisis like famines, floods or earthquakes.
- **Development aid:** This is long-term aid. It involves projects like building schools and hospitals and teaching people how to farm wisely.

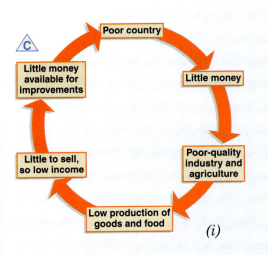

C

(i)

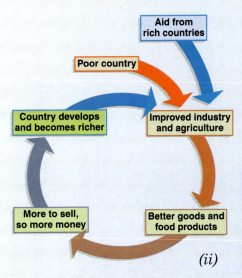

(ii)

(i) How lack of money makes it impossible for a country to develop; (ii) How aid can help a country develop and improve its standard of living

D

Skilled people such as nurses, agriculturalists, managers and teachers who give advice and help to train people

Projects like building new bridges and roads, improving water and electricity supplies, and modernising farming

Equipment for hospitals, farmers, construction and educational material

Food provided either free or at cost price by countries that have a surplus

Money to assist development programmes

Different types of aid

Emergency assistance providing relief to disaster areas

Key points

- There is a flow of aid from the developed countries to the less developed countries.
- Aid is sent from different sources, governments and non-governments.
- Aid can be **emergency** (short-term) or **developmental** (long-term).
- **Tied aid** comes with conditions attached.

Ireland's Aid Programmes

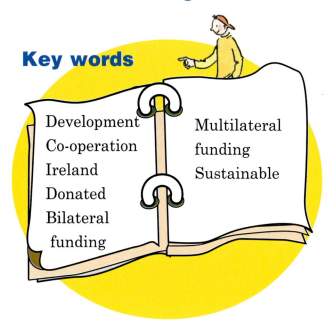

Key words

Development
Co-operation
Ireland
Donated
Bilateral
funding

Multilateral
funding
Sustainable

Multilateral funding means that money is given to organisations such as the United Nations or the Red Cross and they decide how it is spent.

Bilateral funding involves a direct donation of aid from Ireland to another country. Let's look at one of Ireland's bilateral aid programmes in Africa.

Ireland gives aid or help to developing countries through a programme called **Development Co-operation Ireland (DCI)**. In 2003 around €450 million was **donated** (given) from this source.

Ireland also gives aid through organisations such as Concern, GOAL and Self Help. These are called **non governmental organisations** (NGOs).

Development Co-operation Ireland provides **bilateral** and **multilateral funding** for projects in over 75 countries around the world.

Case Study: Ireland's Bilateral Programme in Uganda

Uganda is one of the poorest countries in the world. It has a GNP of $330 per year. Life expectancy is only 43 years. Fifteen years of conflict between 1972 and 1986 led to a weak economy.

In agreement with the Ugandan government, many projects have been funded by the Irish government.

B DCI Spending in Uganda (€mn)			
2001	2002	2003	2004
15.9	35.4	34.2	30.5

A

Countries that receive aid from Ireland. Check on a map of the world (Groundwork 1, page 141)

The countries prioritised in Ireland's Bilateral Aid Programme

Projects funded by the Irish government include:

Health: Support for a national primary healthcare training programme.

Education: Teacher education, educational materials, promoting HIV/AIDS education, adult literacy.

Farming: Advice to farmers about crops, budgets and research.

Non-governmental aid programme in Uganda

One of Ireland's non-governmental aid agencies is **Self Help Development International**. Today it funds projects in six countries in Africa at a cost of $6 million per annum. The idea behind Self Help is that the locals should control how changes are managed. Self Help works with local staff on projects in their areas. Let's look at one example of their work.

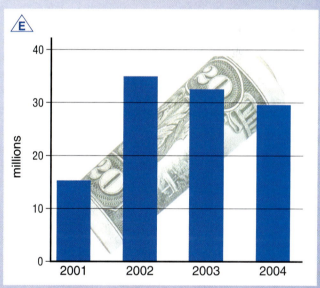

DCI spending in Uganda, 2001-2004 (€mn)

Self Help set up savings and credit schemes that have helped farmers. With her loan from Self Help, Gertrude Batti bought 50 Nera layer hens. During the laying season she gets 23–30 eggs a day. She can sell these eggs for 100 Ugandan shillings per egg. With the money she pays back her loan and has money to spend on her children's education. She has four children, three of whom attend school.

Key points

● Governments and non-government agencies have funded projects in developing countries.
● With the help of DCI and non-governmental organisations like Self Help, countries such as Uganda are making progress.

Factors that Slow Development

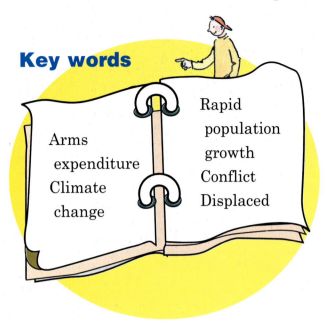

Key words

Arms expenditure
Climate change

Rapid population growth
Conflict
Displaced

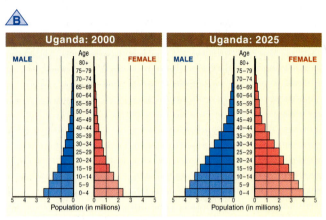

Uganda has a rapidly growing population

You have already learnt that colonisation stopped countries from developing. But there are other factors that slow down development. These include:

● Population growth
● **Arms expenditure** and war
● **Climate change**.

Let's look at how these factors have added to poverty in Uganda.

High population growth

Uganda's population is growing at a rate of 3.4% every year. It has one of the highest fertility rates in sub-Saharan Africa. This stands at seven births per woman. There is a large young population, over 51% are under the age of 14. Uganda also shares a border with several other African countries, many of which are at war. Migrants cross the border adding to the population numbers. With such **rapid population growth**, it is impossible for the economy to improve. The number of people living on less than one dollar a day remained at 9.5 million in 2003, the same as in 1992.

Arms spending and war

Uganda's development is held up because of **conflict**. War is costly both in monetary and human terms.

● Over $100 million is spent each year on arms and war. That is money that could be spent on developing the country.
● Over 1.4 million people have had to leave their homes and farms because of war. When people are **displaced** the land is not cared for and so becomes infertile.
● There is more than $100 million per year in lost production because of war. When people are at war, goods are not being produced.

- Because of war, road and rail lines have been damaged and trade has been disrupted.
- Rates of HIV/AIDS in Uganda's conflict areas are thought to be much higher than other areas of the country. The sick are unable to work.

Climate change

The farming sector is very important to Uganda's economy. It employs up to 83% of the population and accounts for up to 85% of Uganda's export earnings. Crops and animal rearing depend on particular climate conditions. Scientists agree that there has been an increase in world temperatures (global warming). Even small increases can have a very serious effect on farming.

Let's look at the coffee crop and the effect of **climate change** in Uganda. If temperatures rise by two degrees, it would be impossible to grow coffee on the lowlands. It would be too hot. Coffee growing would be confined to the higher land. This would greatly affect the farmers of Uganda and the economy in general.

In many parts of Uganda trees have been cut down to grow more and more crops. Deforestation (cutting down of trees) has led to longer dry seasons. This affects crops. When crops fail, people go hungry. When people are sick they cannot work, they cannot make money and so poverty takes over.

This shows how the economy of developing countries can be at great risk from climate changes.

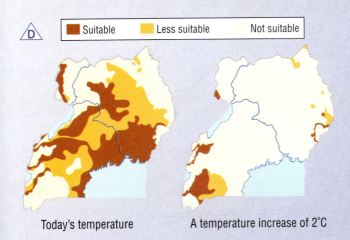

| Suitable | Less suitable | Not suitable |

Today's temperature   A temperature increase of 2°C

Impact of global warming on Robusta coffee in Uganda

Key points

- **Rapid population growth** makes it difficult for Uganda to make progress.
- Spending on arms and the other costs of war also stop Uganda from developing.
- Climate change could affect Uganda's ability to produce crops. This would keep Uganda poor.

Differences within States

Key words

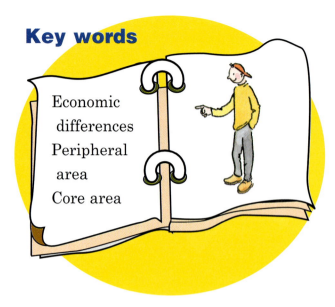

Economic
differences
Peripheral
area
Core area

Wealth is not evenly shared out within countries. Some regions are better off than others. This is true of Ireland as the east of the country is better off than the west. In the case of France, the Paris region is much richer than the Massif Central region.

Let's look first at Ireland.

Economic differences within Ireland

Western region – a peripheral area

- Incomes are 9% below the national average.
- Thin infertile soils, damp cloudy conditions and poorly drained land lower farm outputs.
- Transport links are poorer in the west. Some areas are not connected by rail to the rest of the country.
- Few industries in the West has led to high rates of migration of young people in search of jobs.
- Less income comes from the service industry. The annual tourist spend ($596) is almost half that of the Dublin region.

Eastern region – a core area

- In the Dublin region incomes were 16% above the state average in 1998.
- The deep brown fertile soils of the east provide rich farming land. Farm incomes are high.
- Most of the state's spending on transport has been in this area. The national road and rail links focus on Dublin.
- Most secondary industries have set up here. This brings money and skills to the region.
- The service industry is more developed here. Headquarters of large companies, including banks, are located in the Dublin region.

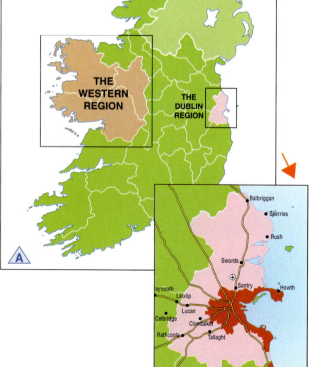

Economic differences within France

Inequalities exist between different parts of France.

Paris Basin – a core area

- The Paris region is a lowland area of fertile soils. It has warm summers (20°C), winter temperatures do not fall below 0°C. Good conditions for crops mean high farm incomes.
- It is the hub of the country's road, rail and air network.
- The River Seine links the region to the port of Le Havre.
- It has a highly skilled workforce. Good reputation of the city of Paris. Headquarters of many successful businesses are located here.
- An area of in-migration with a huge market of over 10 million. It has a young, skilled and highly-educated workforce.

Massif Central – a peripheral area

- Difficult farming conditions: because of steep slopes and poor soils. Winters are very cold and rainfall levels are high. Farm incomes are low.
- There are few natural resources. Supplies of coal have been used up.
- Roads and rail connections are not as well developed as elsewhere in the country. The TGV (rapid train) does not serve the region.
- Low levels of investment in industry and services due to isolation from markets.
- High levels of migration of young people in search of work and education.

Key points

- There are differences in wealth within rich countries.
- The West of Ireland is poorer than the east of Ireland.
- The Paris Basin in France is richer than the Massif Central.

REVISION EXERCISES

Write the answers in your copybook.

1 The poorest countries of the world tend to be:
- In Europe
- In the north of the world
- In the south of the world
- In North America

2 Emergency aid is:
- When the money must be used to buy goods from the donor country
- When money is given by the Irish government to a country to buy farming tools
- When food, blankets and medicine are donated at a time of disaster or famine

3 Which of the following is an example of a non-governmental organisation (NGO)?
- The United Nations
- Concern
- The Brazilian government
- The European Union

4 When teachers and nurses from Ireland volunteer to work in a developing country the type of aid is:
- Tied aid
- Food aid
- Emergency aid
- Development aid

5 Look at the cartoon. Which of these statements is fully correct, according to the advertisements in the shop window?
- Malaysian tea and cotton from India
- Nigerian tin and Chilean copper
- Jamaican bananas and Indian cotton
- Chilean coffee and Jamaican bauxite

6 Look again at the cartoon opposite. The message of the cartoon is:
- Europe does not import food
- We import tin from Kenya
- Countries depend on each other
- Aid should only be for emergencies

7 Look at the pie charts. One shows the economy of a developing country and one shows the economy of a developed country.

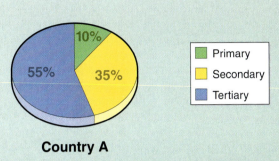

Country A

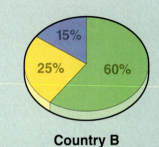

Country B

Which chart shows the economy of a developing country? Explain why you chose this chart.

8 Ireland gives aid to the developing world. Describe **one** example of the type of assistance that Ireland gives to developing countries.

9 Imagine that you are on a visit to an African village with an aid agency. Write a letter to a friend and in it describe **three** things that are different from home. Use the information presented in graphs throughout this chapter as a guide. (Hint: differences in income, health and education.)

10 Explain how the world trade in **coffee** has slowed down the rate at which some of the poorer countries can develop.

11 The chart below shows the main companies that market coffee in the UK.
- (a) What is the overall percentage marketed by Nestlé and Kraft General Foods?
- (b) These companies are called 'multinational' companies. Explain what is meant by this term.
- (c) Why is the domination of the coffee trade by a few big companies unfair to the poor farmer who grows the crops?

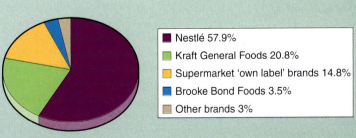

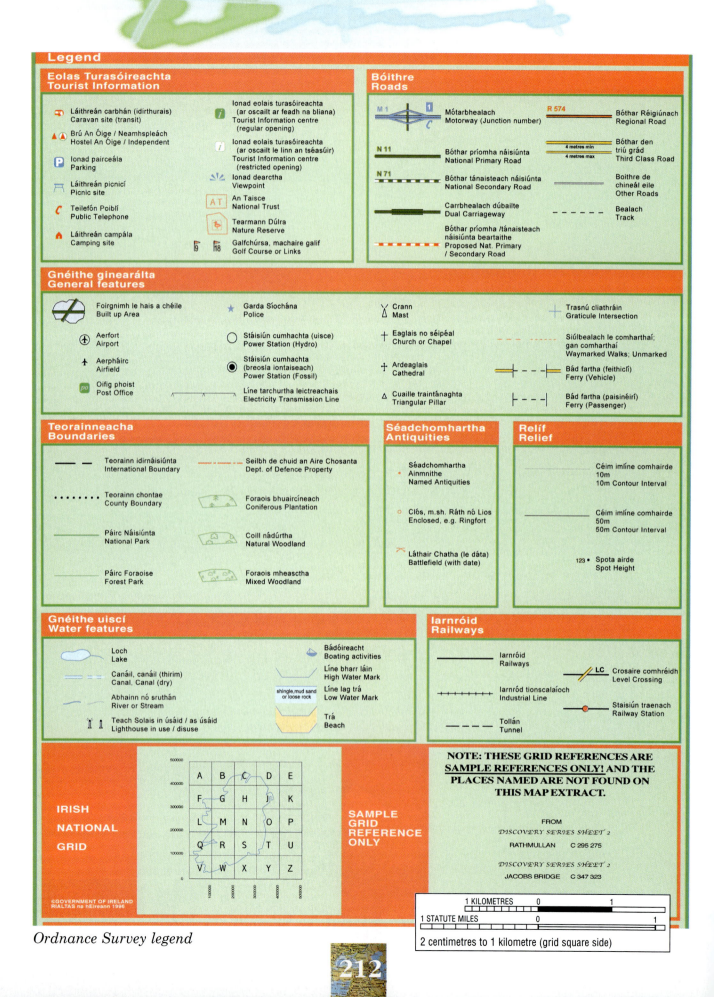

Legend

Eolas Turasóireachta / Tourist Information

- Láithreán carbhán (idirthurais) / Caravan site (transit)
- Brú An Óige / Neamhspleách / Hostel An Óige / Independent
- Ionad pairceála / Parking
- Láithreán picnicí / Picnic site
- Teileofón Poiblí / Public Telephone
- Láithreán campála / Camping site
- Ionad eolais turasóireachta (ar oscailt ar feadh na bliana) / Tourist Information centre (regular opening)
- Ionad eolais turasóireachta (ar oscailt le linn an tséasúir) / Tourist Information centre (restricted opening)
- Ionad dearctha / Viewpoint
- An Taisce / National Trust
- Tearmann Dúlra / Nature Reserve
- Galfchúrsa, machaire galif / Golf Course or Links

Bóithre / Roads

- Mótarbhealach (Junction number) / Motorway (Junction number) — M 1
- Bóthar príomha náisiúnta / National Primary Road — N 11
- Bóthar tánaisteach náisiúnta / National Secondary Road — N 71
- Carrbhealach dúbailte / Dual Carriageway
- Bóthar príomha /tánaisteach náisiúnta beartaithe / Proposed Nat. Primary / Secondary Road
- Bóthar Réigiúnach / Regional Road — R 574
- Bóthar den tríú grád / Third Class Road — 4 metres min / 4 metres max
- Boithre de chineál eile / Other Roads
- Bealach / Track

Gnéithe ginearálta / General features

- Foirgnimh le hais a chéile / Built up Area
- Aerfort / Airport
- Aerpháirc / Airfield
- Oifig phoist / Post Office
- Garda Síochána / Police
- Stáisiún cumhachta (uisce) / Power Station (Hydro)
- Stáisiún cumhachta (breosla iontaiseach) / Power Station (Fossil)
- Líne tarchurtha leictreachais / Electricity Transmission Line
- Crann / Mast
- Eaglais no séipéal / Church or Chapel
- Ardeaglais / Cathedral
- Cuaille traintánaghta / Triangular Pillar
- Trasnú cliathráin / Graticule Intersection
- Siúlbealach le comharthaí; gan comharthaí / Waymarked Walks; Unmarked
- Bád fartha (feithiclí) / Ferry (Vehicle)
- Bád fartha (paisinéirí) / Ferry (Passenger)

Teorainneacha / Boundaries

- Teorainn idirnáisiúnta / International Boundary
- Teorainn chontae / County Boundary
- Páirc Náisiúnta / National Park
- Páirc Foraoise / Forest Park
- Seilbh de chuid an Aire Chosanta / Dept. of Defence Property
- Foraois bhuaircíneach / Coniferous Plantation
- Coill nádúrtha / Natural Woodland
- Foraois mheasctha / Mixed Woodland

Séadchomhartha / Antiquities

- Séadchomhartha Ainmnithe / Named Antiquities
- Clós, m.sh. Ráth nó Lios / Enclosed, e.g. Ringfort
- Láthair Chatha (le dáta) / Battlefield (with date)

Relíf / Relief

- Céim imlíne comhairde 10m / 10m Contour Interval
- Céim imlíne comhairde 50m / 50m Contour Interval
- 123 • Spota airde / Spot Height

Gnéithe uiscí / Water features

- Loch / Lake
- Canáil, canáil (thirim) / Canal, Canal (dry)
- Abhainn nó sruthán / River or Stream
- Teach Solais in úsáid / as úsáid / Lighthouse in use / disuse
- Bádóireacht / Boating activities
- Líne bharr láin / High Water Mark
- Líne lag trá / Low Water Mark
- shingle, mud sand or loose rock
- Trá / Beach

Iarnróid / Railways

- Iarnróid / Railways
- Iarnród tionscalaíoch / Industrial Line
- Tollán / Tunnel
- Crosaire comhréidh / Level Crossing — LC
- Staisiún traenach / Railway Station

Irish National Grid

Sample Grid Reference Only

NOTE: THESE GRID REFERENCES ARE SAMPLE REFERENCES ONLY! AND THE PLACES NAMED ARE NOT FOUND ON THIS MAP EXTRACT.

FROM

DISCOVERY SERIES SHEET 2
RATHMULLAN C 295 275

DISCOVERY SERIES SHEET 2
JACOBS BRIDGE C 347 323

©GOVERNMENT OF IRELAND
RIALTAS na hEireann 1996

1 KILOMETRES 0 1
1 STATUTE MILES 0 1

2 centimetres to 1 kilometre (grid square side)

Ordnance Survey legend

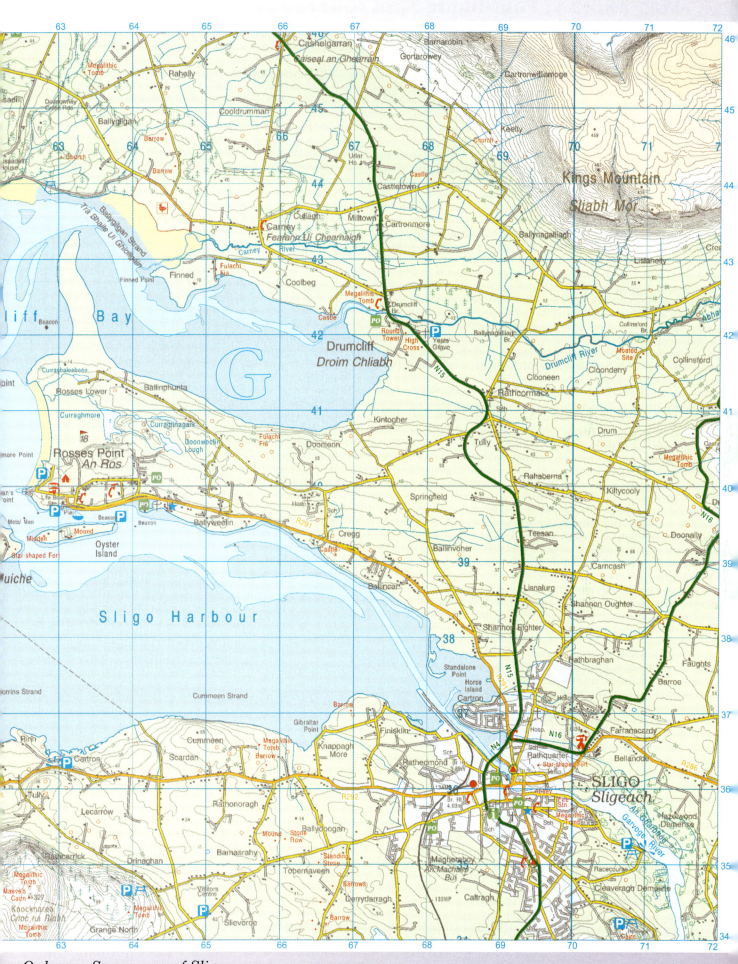

Ordnance Survey map of Sligo

Map Skills 1: Four-figure Grid References

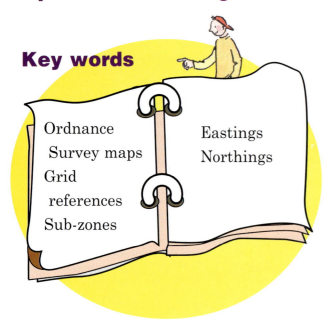

Key words

Ordnance Survey maps
Grid references
Sub-zones

Eastings
Northings

Let's learn map reading skills. The maps you will be studying are made by the **Ordnance Survey** Office (OS maps). We will first look at maps in the Discovery series.

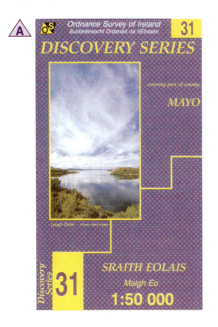

A key or legend helps you to make sense of the different colours and symbols (signs) on the map. Look at the legend on page 212.

Finding places on a map

Let's learn the important skill of reading **grid references**. Grid references help you to find places on a map using letters and numbers.

National grid

There are 25 grid squares covering the map of Ireland. Each of them is named with a letter of the alphabet. These squares are called **sub-zones**.

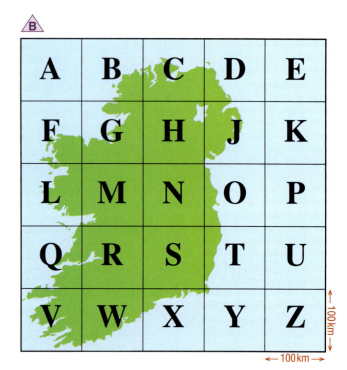

Exercise

Look at map B. Name the **sub-zones** where you would find the following counties: Galway, Donegal and Wexford.

Sub-zones – a closer look

Sub-zones are big areas. You might need to find a town in a sub-zone. To do this you would need a more detailed reference point for the town.

Each sub-zone is divided into 1,000 smaller squares. The lines are drawn in blue and each line is numbered (see diagram C).

Look at diagram 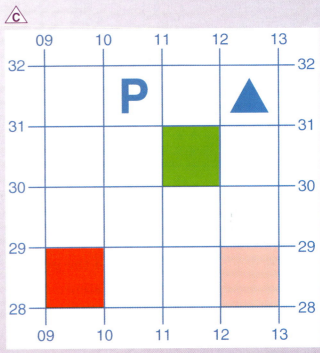. This is part of sub-zone P. Let's find the grid reference for the red square in this sub-zone. Follow these steps:

1. First give the number of the line running down the left side of the square. This number is 09.
2. Then give the number of the line running along the bottom of the square. This number is 28.
3. Put the numbers together: 09 28.
4. Put the letter of the sub-zone before this number: P 09 28.

> Remember to give the grid reference numbers in the following order:
> First, give the number across the top.
> Then give the number along the side.

Exercises

Look at diagram 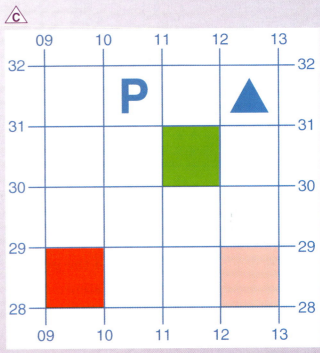. Do the following examples. Put the letter of the sub-zone (in this case P) before each set of figures.

1. Give the grid reference for the green square.
2. Give the grid reference for the pink square.
3. Give the grid reference for the square with the triangle.

Sub-zone P

Naming the grid lines

● The blue grid lines running down the map are called **Eastings**.
● The blue grid lines running across the map are called **Northings**.

Key points

● **Grid references** describe where places are on a map.
● The grid reference is given in the following order: **sub-zone, Eastings, Northings**.

Map Skills 2: More about Grid References

Key words

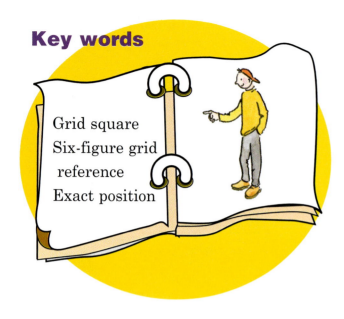

Grid square
Six-figure grid
 reference
Exact position

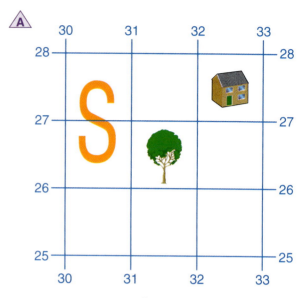

In the last section you learnt to *give* a grid reference for a place. Let's learn to *find* places using a grid reference.

Look at diagram △A. Let's find the **grid square** S 32 27. Follow these steps:

1. Go across the top until you find the numbered line 32.
2. Now go up the side of the diagram until you find the numbered line 27.
3. Follow line 32 down the map and line 27 across the map. The two lines meet to form an L around the grid square S 32 27.

Six-figure grid references

Let's learn to read **six-figure grid references**. This is useful if you want to find a more **exact position** on a map.

Look at diagram △C. Let's find the exact position of the red dot.

1. Start off with the line to the left. This is 04.
2. Move across the square to the position of the red dot. In your head divide the square into tenths as shown in the example. The red dot is five-tenths of the way into the square. Write the number 5 after the 04 = 045. This is the first part of the grid reference.
3. To find the second part of the grid reference look at the line on the bottom it is 72.
4. Divide the square into tenths. The red dot is three-tenths of the way up from 72. Write the number 3 after the 72 = 723. This is the second part of the grid reference.
5. Put the two parts together: 045 723.
6. Put the sub-zone letter before this.
7. The grid references of the red dot is Z 045 723.

B

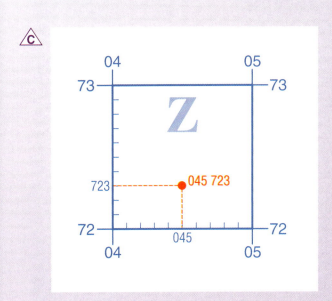

C

Exercises

Look at map B.

Name the features at the following grid references.

(a) G 681 441

(b) G 658 434

(c) G 631 444

Exercises

1. Look at diagram A.

   (a) What would you find in the grid square S 32 27?

   (b) What would you find in the grid square S 31 26?

2. Look at map B.

   (a) What would you find in grid square G 70 44?

   (b) What would you find in grid square G 64 43?

   (c) What would you find in grid square G 63 45?

Key points

- A **six-figure grid reference** gives a more **exact position** on the map.

Map Skills 3: Direction

Key words

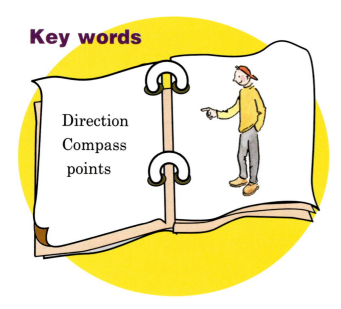

Direction
Compass
points

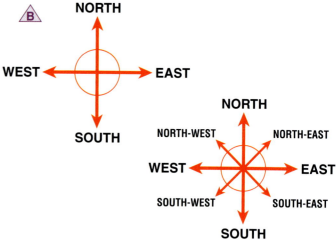

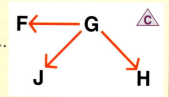

Let's learn about **direction** on a map. To do this you must use the **points** on a **compass**.

There are four main points on the compass. These are north, south, east and west. Between the four main points there are four other points. Look at diagram B.

When asked to describe the **direction** on a map you must state the direction you need to go to get to a place. Look at the following examples. Note that it is really important to know your *starting point*. In this case it is A.

● The direction from A to B is east.
● The direction from A to C is north-east.
● The direction from A to D is south.

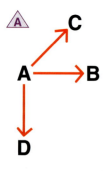

Exercise

Look at diagram C. Your starting point is G.

(a) What direction would you need to go to get to F?

(b) What direction would you need to go to get to H?

(c) What direction would you need to go to get to J?

To name the direction of one town from another as the crow flies you take the most direct path between the two places.

Exercise

Look at the OS map of Sligo on page 213 and answer the questions.

1. What is the direction as the crow flies from Sligo G 69 36 to Rosses Point G 64 40?

2. What is the direction as the crow flies from Rosses Point G 64 40 back to Sligo G 69 36?

3. What is the direction as the crow flies from Sligo G 69 36 to Kings Mountain G 70 44?

4. What is the direction as the crow flies from Cartron G 63 36 to Sligo?

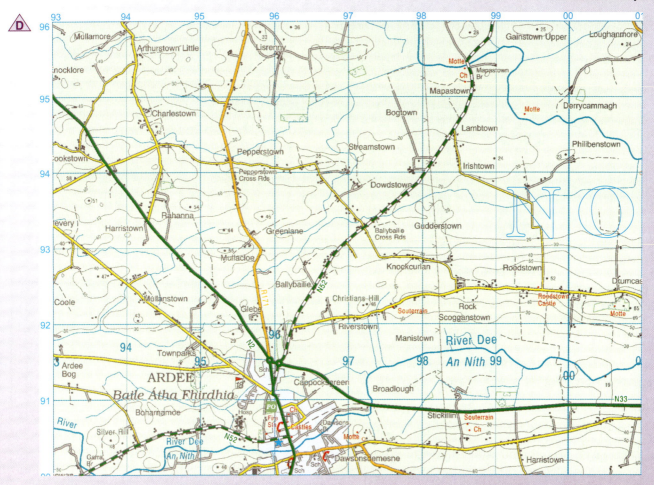

Direction followed by a road

To name the direction followed by a particular road, find your starting point and state the direction you need to take to get from one end of the road to another.

North on OS maps

The top of a map is always north. The grid lines that run down the map (Eastings) run from north at the top of the map to south at the bottom of the map.

> North? Which way?

Exercise

Look at the OS map **D** of Ardee. What direction would you be travelling if you were driving along the following roads?

(a) The N2 (shown as a green line) from Harristown N 944 934 to Ardee N 962 905.

(b) The N52 (shown as a green and white line) from Dowdstown N 979 939 to Ardee N 962 905.

(c) The R171 (shown in orange) from where it comes onto the map at N 955 940 to Ardee N 962 905.

Key points

● **Direction** refers to the direction you need to take from your starting point to get to another place.

● The top of the map is always north.

Maps Skills 4: Symbols on Maps

Key words

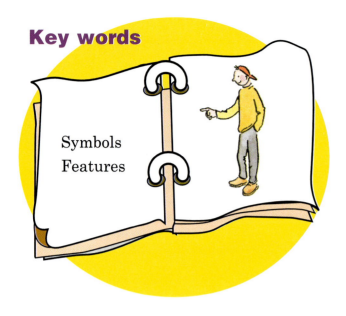

Symbols
Features

Exercise

Look at the OS map of Ardee on p. 237 and use the legend to find the meaning of the symbol.

Make a copy of this table. Fill in the spaces. The first one has been done for you.

Grid reference	Symbol	Meaning
N 949 869	po	A post office
N 961 896		
N 987 858		
O 014 865		

Symbols are used to represent **features** on maps. This is necessary because quite a lot of information has to be given on a map. Symbols save space and make things clear. Look at the following symbols. Can you say what they stand for?

Tourism on OS maps

Tourist attractions and facilities can be read on a map. Symbols to show some of these can be seen in diagram B.

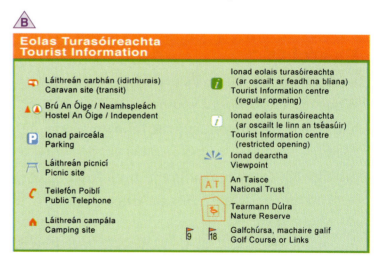

Let's look at maps. The main symbols are shown on the legend on page 212. Small drawings, shortened words, lines or colour are used. The symbols are explained in the key or legend that comes with the map.

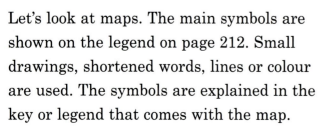

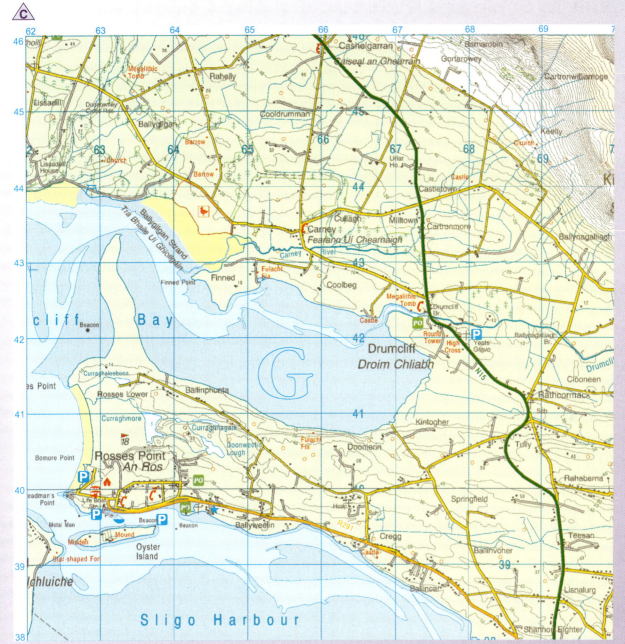

Map ⚠C *shows some of the reasons why tourists visit an area.*

Exercise

Look at Map ⚠C and find features at the following grid references:

(a) G 667 423

(b) G 626 440

(c) G 631 398

(d) G 629 400

Key points

● **Symbols** on maps represent features or activities in an area.

Map Skills 5: Scale and Distance on a Map

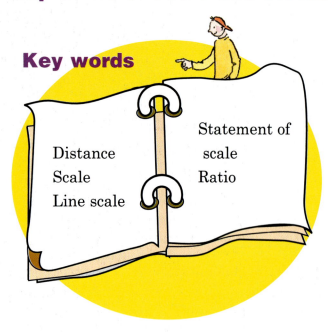

Key words

Distance
Scale
Line scale
Statement of scale
Ratio

A map is a scaled drawing of real **distance**. The **scale** of the map can be shown in three different ways:

1. **Line scale:** You can show the scale of the map on a line. Look at diagram A.

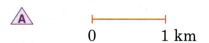

(A)   0   1 km

This line measures 2 cm on the map but it represents a distance of 1 km on the ground.

2. **Statement of scale:** You can show scale by stating the scale. You can say that 2 cm on the map stands for 1 km in real distance on the ground.

3. **Ratio:** You can show scale by writing the scale as a ratio. This is also called a representative fraction (RF).

Ratio is a way of comparing the size of the map with the size of the real place. It looks like this: 1:50,000. This means that the map is 50,000 times smaller than the real area covered. It means, for example, that every 1 cm on the map stands for 50,000 cm

on the ground (50,000 cm = 0.5 km).

1 cm = 0.5 km

2 cm = 1 km

Look at the scale on p. 212.

Measuring the area covered by the map

Area is the length multiplied by the width. Each square on an OS map measures 1 km^2. To measure the area of the map, you count up the number of grid squares across the top and the number of grid squares down the side of the map. Multiply them. This gives you the area in kilometres squared (km^2).

Measuring distance on the map

You can measure how far one place is from another by using a strip of paper. Let's first take the shortest route between two places, the straight line.

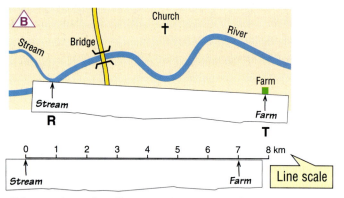

Measuring the distance in a straight line between two places

1. Place a straight-edged piece of paper between the two end points (e.g. R and T).
2. Mark the piece of paper at these end points.
3. Place the piece of paper along the **line scale** placing the first mark under the 0 on the line. You will now find out the actual distance on the ground between R and T. In the above example it is 7 km.

Measuring distance on a curved line

To find the distance along a curved line, measure the distance in straight-line sections. Look at diagram △C. You are asked to measure the length of the river on the map.

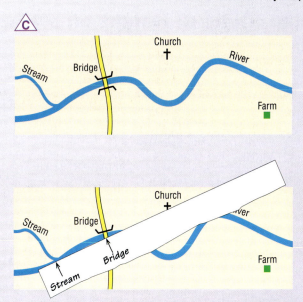

1. Get a straight-edged piece of paper.

2. Place the paper on the map and line it up along the first straight section on the river. Mark this point on the edge of the paper.

3. Turn the paper until you can line it up along the next straight section. Mark this on your straight-edged piece of paper.

4. Twist the paper until it lies along the next straight section. Mark the paper again.

5. Keep doing this, twisting and marking the paper until you reach the end of the line being measured.

6. As before, line up the strip of paper along the line scale and find out the distance in kilometres (km).

Exercises

1. What is the area covered by the Sligo map on p. 213?

2. Look at the Sligo extract.

   (a) Calculate the straight-line distance between the following two points: the junction of the N4 and N16 roads at G 691 367 to where it leaves the map at G 720 394.

   (b) Calculate the distance as the road goes (curved-line distance) between the same two points.

   (c) Give one reason why the road does not follow the straight line route between these two points.

Key points

● Use the line scale given in the legend to measure real distance on the ground.

Map Skills 6: Height on a Map

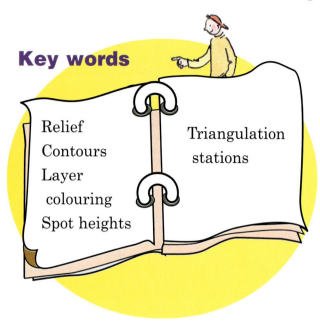

Key words

Relief
Contours
Layer colouring
Spot heights

Triangulation stations

Land is rarely flat. There are hills, mountains and valleys. Sometimes slopes are gentle, other times they are steep. The lie of the land, its shape, is called the **relief** of the land. This can be shown on a map.

One of the difficulties in showing relief on a map is that you are trying to show the height and shape of features on the flat surface of a map. Map makers have four ways of showing the relief of the land:

- **Contours**
- **Layer colouring**
- **Spot heights**
- **Triangulation stations**.

Contours

Contours are lines on a map that join places that lie at the same height above sea level. The contour lines are like light pencil lines. The height, in metres, is usually written along the line. You may need to follow the line along the map to find the number.

There is a 10-metre difference between one contour and the one beside it. We call this a contour interval of 10 metres. So although the map is flat, the contours can be read and the shape and height of the land can be imagined.

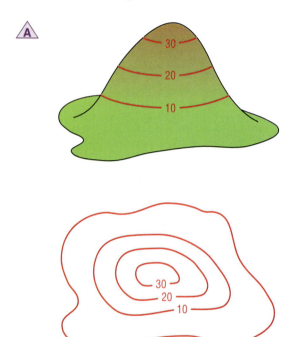

A

Contour interval = 10 metres

Example of contours on map **B**.
G 683 450 = 90 metres
G 693 436 = 50 metres

Exercise

Look at map **B**. What are the contour heights at the following grid references?
G 687 435
G 717 458
G 717 433
G 686 450

Layer colouring

Height is shown on maps in metres. Areas of different heights are shown in different colours. A map key will show you the meaning of the colours. Look at diagram C.

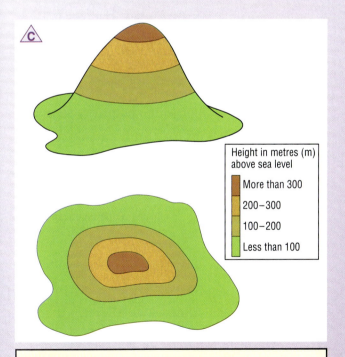

Height in metres (m) above sea level

- More than 300
- 200–300
- 100–200
- Less than 100

Exercise

How many tints of layer colouring are used on the Sligo map on page 213?

Spot heights

A spot height is marked on the map with a spot and a figure that shows the exact height of the land at that spot. The height is always shown in metres. **Triangulation stations** show height in the same way. They are drawn as a dot inside a triangle. You will find a triangulation station on the top of a hill or a mountain. Look at diagram D.

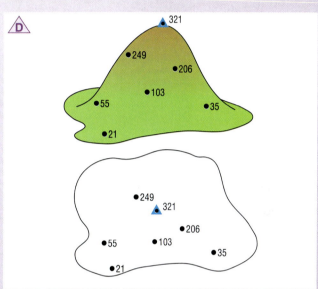

Examples from the Sligo map on page 213 of spot heights:

G 642 408

G 709 439

G 646 356

Exercises

Look at map B.

1. What is the height of Kings' Mountain (G 704 443)?

2. How high up is the Church at G 688 445?

Key points

- **Relief** is the lie of the land
- It can be shown on a map using colour, **contours**, **spot heights** and **triangulation stations**.

Map Skills 7: Slopes

Key words

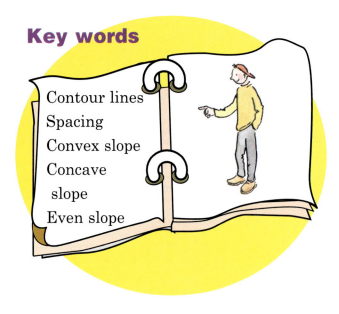

Contour lines
Spacing
Convex slope
Concave
  slope
Even slope

Contour lines on the map represent the height of the land. Contours show more than just the height of the land. The shape or pattern of the contours also tells something about the shape of the land. Look at the diagrams.

The **spacing** of the contours tells something about the slope of the land.

- Contour lines close together = land is steep.
- Contour lines more spaced out = slope is gentler.

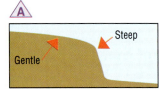

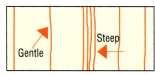

Convex and concave slopes

Sometimes hills and mountains have sections that are steep and sections that are gentle. Picture yourself with a map, planning your routeway up a hill. You have a decision to make. From which side will you climb the hill? Which slope looks better to you?

- One starting point offers the steep section first, followed by a more gentle section. This is a **convex slope** (diagram ◮B◮ (2)).
- Another starting point offers the gentle section first followed by the steep section. This is a **concave slope**. (diagram ◮B◮ (3)).
- Yet another route up the mountain offers a more **even slope**. It does not have steep or gentle sections. The slope stays the same. (diagram ◮B◮ (1)).

By looking at the contour spacing you can work out which kind of climb suits you best. This is because contour spacing shows the shape.

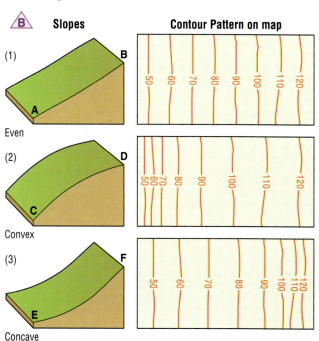

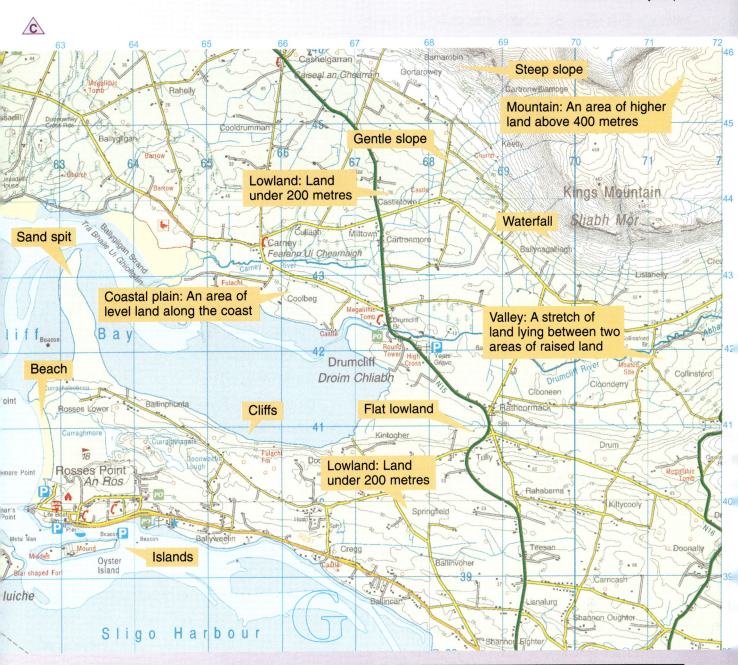

Map C: Ordnance Survey extract showing the Sligo / Drumcliff area with landscape features labelled:
- **Steep slope**
- **Mountain: An area of higher land above 400 metres**
- **Gentle slope**
- **Lowland: Land under 200 metres**
- **Waterfall**
- **Sand spit**
- **Coastal plain: An area of level land along the coast**
- **Valley: A stretch of land lying between two areas of raised land**
- **Beach**
- **Cliffs**
- **Flat lowland**
- **Lowland: Land under 200 metres**
- **Islands**

Landscape features on a map

You have already studied different types of landscapes. Landscapes are shaped by weathering, by rivers, by the sea and by ice. An OS map shows landscape features by using different colours and different contour shapes. Look closely at the OS extract (diagram C) which shows how these features are depicted.

Drainage

Drainage is the word for rivers and lakes on the landscape. These are shown in blue on the map.

Exercise

Look at map C. Using grid references find the location of:

- A river
- One lake
- One waterfall
- A sea cliff
- Inland cliffs.

Key points

- From the colours and line shapes on a map, you can work out the landscape features in an area.

Map Skills 8: Communications

Key words

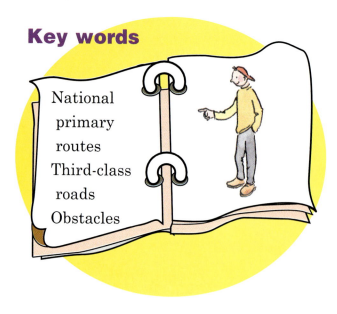

National primary routes
Third-class roads
Obstacles

- Mountains
- Steep slopes
- Marsh or boggy land
- Easily flooded land such as the area close to a river.

Communications are the road, rail, canal, sea and air links.

Roads

Look back at the OS legend on page 212. Notice that there are different types or classes of roads.

The most important roads are the **national primary roads**. They carry the most traffic and link large settlements. Less important roads, such as **third-class roads**, link small settlements such as isolated farmhouses.

Where to build a road?

Roads exist to take traffic from one place to another. The aim is to take the most direct route – a straight line if possible – as this is the shortest route.

It is not always possible for a road to run in a straight line between two places. It has to avoid **obstacles** that lie in its way. The following features on landscapes are obstacles:

For these reasons you are more likely to find roads:

- Along the line of a valley. The road follows the easy route that the river had already carved out between the hills and the mountains.
- Running through gaps in the mountains. Roads avoid steep slopes.
- Running along the lowlands where they link most settlements and are not so expensive to build.
- Running across steep slopes in a zigzag way.

Look at the map of Kenmare (diagram A). It shows how the roads are linked to or related to the relief.

- The R569 (V 915 710) runs across the lowland in long straight sections.
- The N71 (V 920 697) follows the river valley. This is the easiest route as it avoids the mountains.
- The third-class road bends to follow a gap through the mountain at V 895 635.
- The R571 follows the line of the coast.

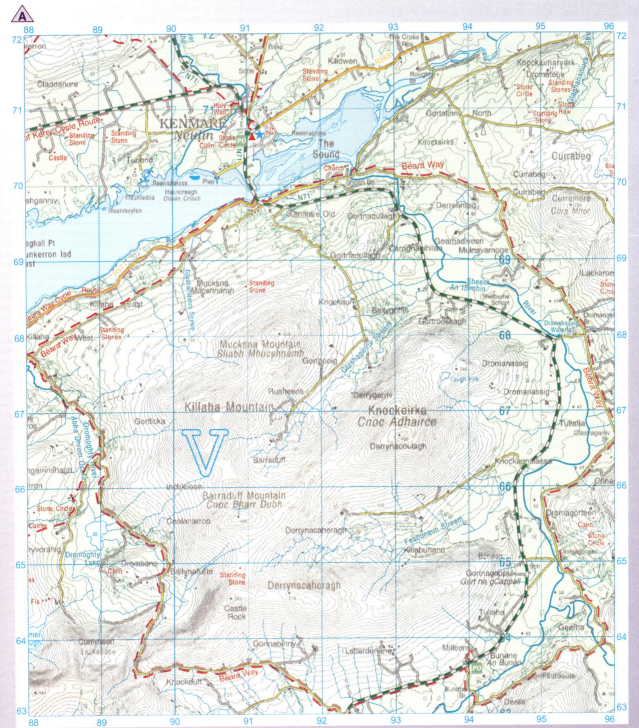

OS map of Kenmare

Exercise

Using grid references point out and explain three places where the roads are related to the relief of the land.

Key points

- There are different classes of roads. The map shows these differences by using coloured lines and names.
- Most roads follow easy routeways such as lowlands and valleys.

Map Skills 9: Settlement on Maps

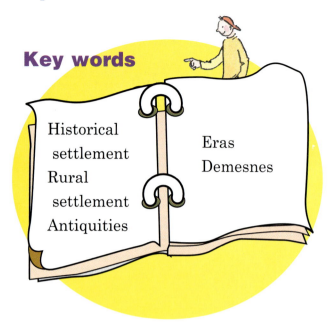

Key words

Historical settlement

Rural settlement

Antiquities

Eras

Demesnes

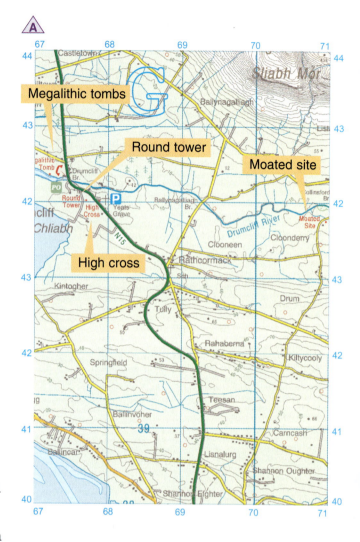

Settlement refers to buildings on the landscape. Let's discover how settlement is shown on a map. We will look at:

- **Historical settlement**
- **Rural settlement**.

Historical settlement

Evidence of historical settlement is shown on the map with a red dot or other symbol. The name is written in red. Look at the legend on p. 212 and read the piece on **antiquities**. Antiquities are historical sites on the landscape.

These historical sites date from different **eras**. There is evidence from:

- *Pre-Christian times*: Megalithic tombs, cairns, standing stones (Neolithic) ring forts, fulacht fia (Bronze and Iron Age).
- *Early Christian settlements*: Holy wells, graveyards.
- *Medieval settlements*: Norman castles, priories.
- *Plantation settlements*: **Demesnes**.

Look at map ⬛Ⓐ and notice the spread of these sites on the landscape.

Exercise

Look at the OS map extract of Sligo on p. 213. Using grid references find the location of:

- a castle
- a fulacht fia
- a ringfort.

Rural settlements

Rural settlements are settlements outside towns and cities. Settlement is shown on the map with black dots.

Map shows where these settlements have been built. From the map you will be able to see:

- The pattern of rural settlement.
- The type of land the settlements have been built on.

Pattern of settlement on maps

Settlement pattern refers to the shape that houses make on the landscape. You find:

- Nucleated settlements
- Linear settlements
- Dispersed settlements
- No settlement.

Where you find rural settlements

Settlements are found in places where people can make a living. People avoid:

- High land (land above 200 metres)
- Steep slopes
- Badly drained land or land that is easily flooded.

On the other hand you will find high densities of rural settlement:

- On well drained lowland
- Close to main roads
- In river valleys
- On south-facing slopes where there is more sunshine and warmth than north-facing slopes.

Exercises

1. Look at map A. Give two pieces of evidence to show that people have settled in this area for a long time.
2. Look at the Kenmare map on p. 229. Give examples of where the three patterns of settlement can be found.
3. Give two pieces of evidence from map B to show how settlement is linked to landscape.

Include grid references for each answer.

Key points

- Maps show the history of settlement in an area.
- Maps show the pattern of **rural settlement**.
- Maps show how settlement is linked to landscape.

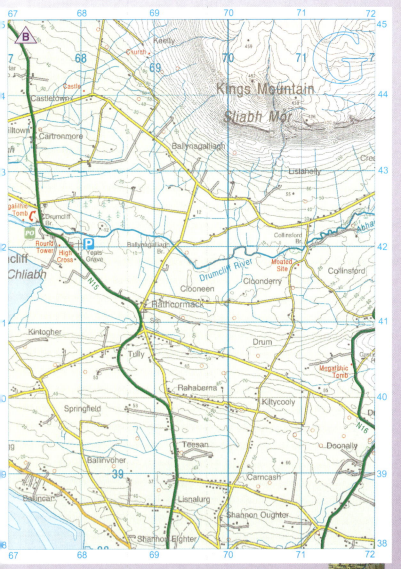

Map Skills 10: Urban Settlement and Tourism on Maps

Key words

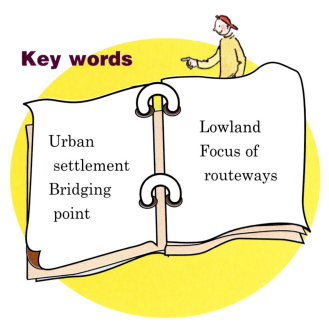

Urban settlement
Bridging point

Lowland
Focus of routeways

Urban settlement refers to towns and cities. Look at map **A** and notice that the town of Sligo is shaded in grey to show the built up area.

Map **A** shows us where a town or city is set up on the landscape. By reading the evidence of the map we can work out why a particular place was chosen for a town. In general, towns develop near rivers, on lowlands and where there are good transport links.

Rivers

The town was set up close to a river. The river is good for a water supply and for transport.

The town was set up at a place along the river where it was easy to cross. We call this a **bridging point**. Maybe the river was narrower at this point and it was easy to build a bridge.

Lowland

The town was set up on **lowland**. It was easy to build roads and houses on lowland.

Transport links

The town was set up where several roads met. We call this a **focus of routeways**. The town was set up near the mouth of a river and so grew as a port.

Urban functions

Function has to do with the reason for something. Towns exist for a number of reasons. Map **B** shows some of the functions.

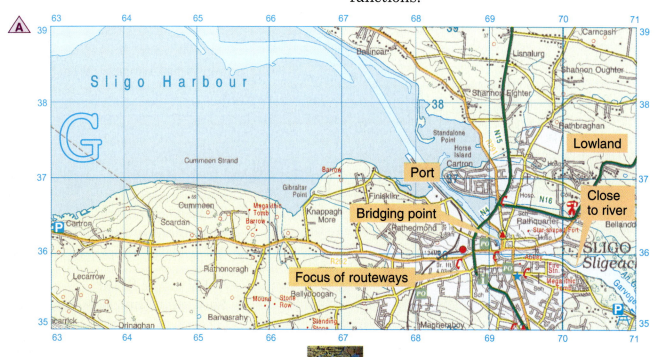

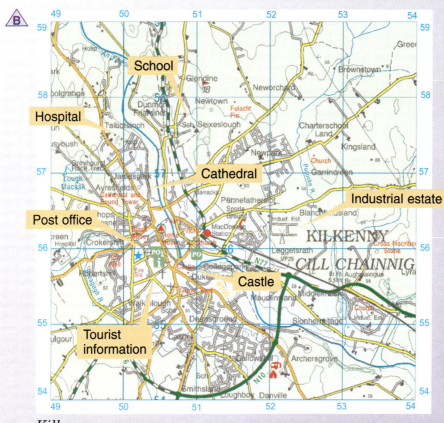

Kilkenny

Functions of a town	Map evidence
Defence	Castle or town wall
Religious	Cathedral, church, abbey or monastery
Industrial	Factory or industrial estate
Services	Post office, tourist information, hospital
Education	Schools or colleges
Transport	Roads, railway station
Housing	Houses

Placenames

Exercises

1. Give three reasons why the site of Kilkenny was chosen for a town.
2. Name three functions found in Kilkenny.
3. Look at the Sligo map (p. 213). Using grid references, find three placenames beginning with the root words in the list in diagram C.

Key points

● Maps show why a town develops in a particular place.
● Maps show evidence of the functions of towns.
● Maps show evidence of tourist attractions.
● Maps give evidence of land use in symbols and placenames.

Map Skills 11: Drawing Sketch Maps from OS Maps

Key words

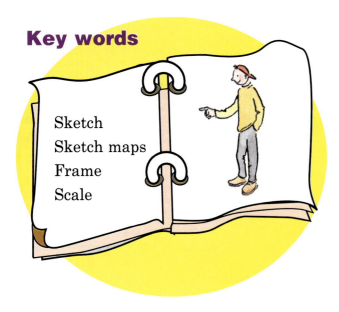

Sketch
Sketch maps
Frame
Scale

Let's learn how to copy or **sketch** a map from an OS map. You will need to mark in and name a number of features. Follow the steps below.

1 Frame

Start by drawing a frame for your sketch. It should be smaller than the map to be copied but have the same shape.

2 Copying

The aim in drawing **sketch maps** is to copy things on the map and put them into your own sketch. To do this correctly you have to draw guidelines on your sketch.

When you were younger, you may have learned how to copy an image of a cat. You drew a grid of squares over the original picture and then drew a grid of squares on the page onto which you were going to make your own sketch. Then you copied what was in each square.

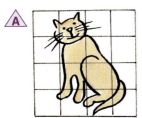

Do exactly the same when copying maps.

- Draw a grid of nine squares over the original map.
- Then draw a grid of nine squares into the **frame** you have already drawn.
- Use the grid squares to guide you where to put things.
- Use a symbol to show the feature e.g. a cross to show a church.
- Label the features carefully.

When you have sketched and labelled the features:

- Put a title on your sketch. This could be the name of the biggest town on the map.
- Put in a direction arrow pointing north.
- Write out your own legend or key to the features you have put in on the map.

Look at the example.

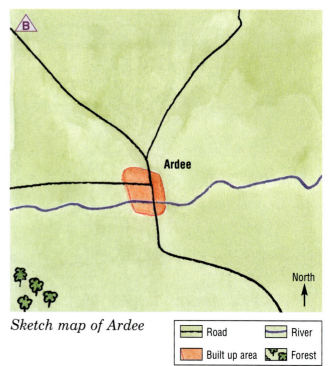

Sketch map of Ardee

	Road		River
	Built up area		Forest

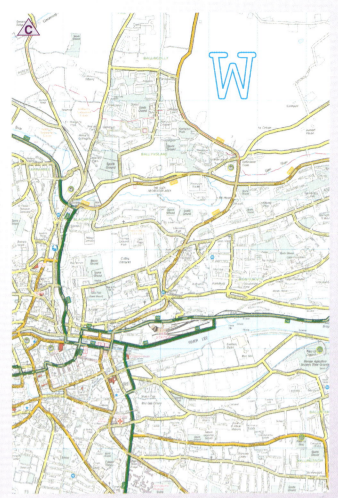

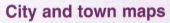

Map 1 Scale 1:20,000

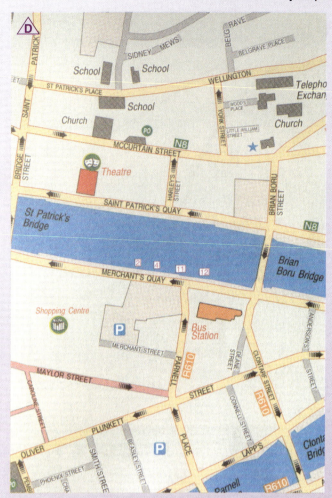

Map 2 Scale 1:8,500

City and town maps

So far we have studied OS maps which are drawn to a **scale** of 1:50,000. Let's look at different map scales.

Scale and detail

Look at the two maps ⚠C and ⚠D. Both show part of the city of Cork.

Notice the differences

Map ⚠C covers a bigger area of ground. Map ⚠D shows more detail. It is a 'close up' view.

As with OS maps in the Discovery series, they have a legend that shows what the different colours and symbols mean.

Street names and named services

Detailed maps of towns show the street names and the names of some buildings and services.

This kind of information is useful to a number of people including visitors to the town.

Key points

- Keep **sketch maps** simple.
- Maps of towns are drawn to a **scale** bigger than 1:50,000. Therefore they are more detailed.

REVISION EXERCISES

Write the answers in your copybook.

1 The contour pattern shows an example of which slope?
   ● Concave slope
   ● Convex slope
   ● Vertical slope
   ● Even slope

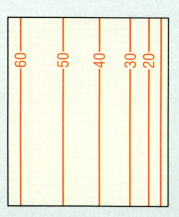

2 The area (in square kilometres) shown on the map of Ardee opposite is about:
   ● 144 sq. km
   ● 110 sq. km
   ● 48 sq. km
   ● 228 sq. km

3 Using the OS map of Ardee opposite, match each letter in column X with the number of its pair in column Y.

X	Y	ANSWER
A N 947 952	1 Between 40m–50m	A =
B O 017 844	2 Spot height 42m	B =
C N 939 8911	3 Contour line 30m	C =
D O 011 926	4 Spot height 127m	D =

4 On the OS map of Ardee opposite, what is the distance, in kilometres, along the N2 road from the junction with the third-class road (N 994 848) to the junction with the R170 road (N 963 903)?

5 Draw a sketch map of the area shown on the map of Sligo on p. 213. Mark on and identify: a post office, the coast, a beach, a megalithic tomb, a picnic site, the town of Sligo, a railway station. Remember to include a frame, title and key.

6 Look at the map of Ardee opposite. Compared with other areas shown on the map, there are less houses around the River Dee between N 98 91 and O 01 91. Using information from the map give **two** reasons why this area has a lower population.

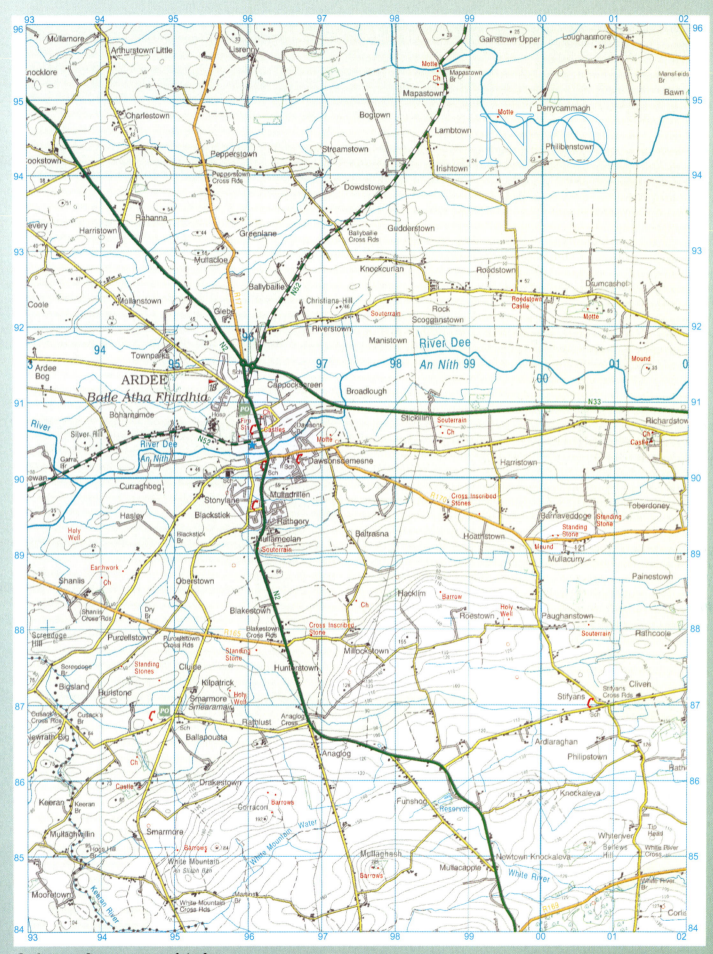

Ordnance Survey map of Ardee

Key words

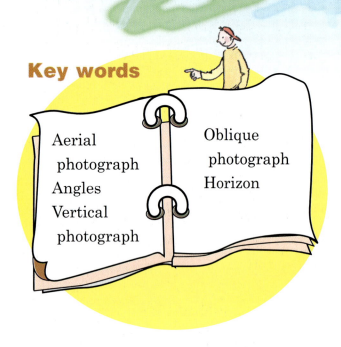

Aerial photograph
Angles
Vertical photograph

Oblique photograph
Horizon

An **aerial photograph** is a *bird's-eye* view of a landscape. Small planes or helicopters fly over the landscape and photographs of the ground underneath are taken from the air.

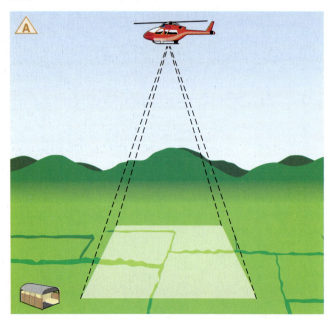

Looking at landscapes from the air

Aerial photographs offer lots of clues or evidence about a landscape. When looking at a photograph, imagine you are a detective searching a scene for evidence.

Types of photographs

Photographs can be taken from different **angles**.

● **Vertical photograph**: This photograph is taken when the camera is facing directly downwards.

● **Oblique photograph**: This photograph is taken when the camera is pointed at a slant or angle across the landscape. In an oblique shot, features nearer to the bottom of the photograph appear larger.

Vertical photograph

Look at photograph B.

● Roof tops are clearly seen.
● Sides of buildings are not seen.
● Features are scaled down equally.
● The **horizon** is not seen.

High oblique photograph

Look at photograph C.

- The horizon can be seen.
- Features close to the bottom of the photograph appear to be bigger.

Low oblique photograph

Look at photograph D.

- Features are larger than in the high oblique photograph.
- The horizon cannot be seen.
- The sides of buildings can be seen clearly.

Naming where something is in a photograph

You need to be able to name where things are in a photograph. When looking at a **vertical photograph** imagine a grid of nine boxes is placed on top of it. The boxes are named after the compass points.

The symbol showing where north is will be drawn on the vertical photograph in the Junior Certificate exam. You describe the location by referring to the compass points.

- There is a roundabout in the south-west of the photograph.
- There are industrial buildings in the south-east of the photograph.

Key points

- **Aerial photographs** are taken from the air. Some are **vertical** and some are **oblique**.
- In an **oblique photograph** the foreground is nearer to the camera, and is at the bottom of the photograph.
- The background appears at the top of the photograph and is furthest from the camera in an **oblique**

Location on an Aerial Photograph

Key words

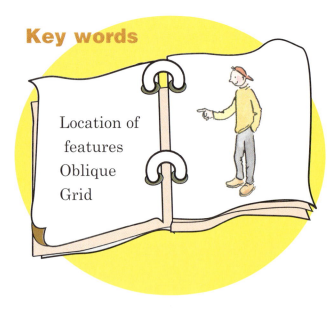

Location of
features
Oblique
Grid

You learnt about naming the **location of features** on a vertical photograph in the last section. When the photograph is an **oblique** one (taken at an angle) you locate a place by using the names listed in the **grid** in photograph A.

Examples of naming locations on an oblique photograph

● The fishing boat beside the pier is in the centre of the photograph.
● There is a small island in the right background.

Look at the lists of features that appear next to photograph B. You need to be able to recognise these features on photographs. Read through the lists and examine the photograph to see what these features look like on a photograph.

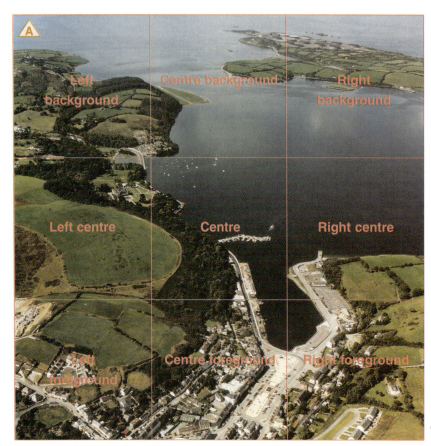

Oblique photograph of Bantry Bay

Aerial view of Howth

Features
1 Pier
2 Railway
3 Car park
4 Marina
5 Railway station
6 Beach
7 Fishing boats

Practise naming the location of the features
in the photograph.

Key points

● To locate a **feature** on a photograph draw a
  grid of nine squares on an aerial photograph.
● Name the **grid** squares.
● Then state where the feature is on the **grid**.

Drawing a Sketch of a Photograph

Key words

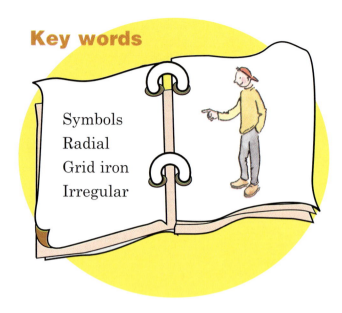

Symbols
Radial
Grid iron
Irregular

Let's learn to make sketch maps of photographs. A sketch map is a ground plan of a photograph.

Follow these steps:

1. Draw a grid of nine squares on a photograph.
2. Draw a nine-box grid in your copybook. It must be the same shape as the grid on the photograph, but it needn't be the same size as the grid.
3. Copy features as asked onto your sketch using the boxes as a guide.
4. When marking them on your sketch, don't draw a picture of the feature. Copy the shape or use **symbols** to show the feature.
5. Include a key that explains the symbols that you used when marking in the features.
6. Use a pencil.
7. Name the feature using clear print.
8. Put a title on your sketch.

Look at the example that has been drawn for you.

Note: Keep the sketch very simple. This is not a test of your artwork. Being clear is the most important point. Being quick, but careful, is also important.

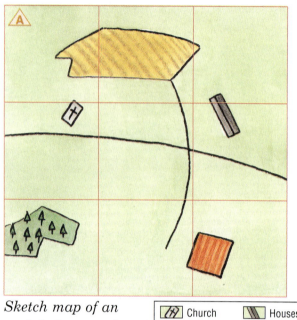

Sketch map of an aerial photograph

Symbol	Feature	Symbol	Feature
	Church		Houses
	Crops		Trees
	Timber yard		

Naming and marking functions on a sketch map

You may be asked to mark and name the functions of a town on a sketch map. Look at the following list of activities or services in a town.

Name of service	Examples
Education	School
Commercial	Shop
Religious	Church
Medical	Hospital or surgery
Transport	Bus or rail service
Residential	Houses
Industrial	Factory
Information	A tourist centre
Administration	A post office

Look at photograph of Tullamore and find examples of some of the activities that are carried on in this town.

Aerial view of Tullamore

Street patterns of towns

The layout of streets in a town can form a pattern. This can be seen on a photograph and copied onto a sketch. It can be:

- **Radial** pattern: Streets fan out from a central square or diamond.
- **Grid iron** or planned pattern: Streets are straight and run at a right angle into each other.
- **Irregular** pattern: Streets vary in width or streets are winding or twisting through the town.

Exercise

Draw a simple sketch of Tullamore and show and name three services. Sketch the layout of the streets.

Key points

- Keep sketch maps simple. Use a pencil or light colouring pencil.
- Do not draw a picture of a feature on a sketch. Draw the shape or use a **symbol**.

Choosing a Site on a Photograph

Key words

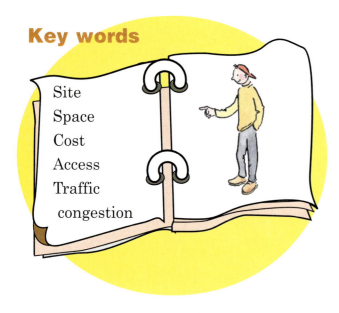

Site
Space
Cost
Access
Traffic
congestion

Let's look at photographs to find out if a **site** is suitable for a particular activity. The site is a plot of ground. You could be asked to choose:

- A site for a new factory.
- A site for a shopping centre.
- A site for a new house.

Before looking at the photograph think about what the site needs to be suitable for this activity. Having decided this, look at the photograph for a site that satisfies these needs. It is *always* the case that:

- It will need enough **space** for the building or activity. Remember to consider the **cost** of this space. Land is cheaper as you move towards the edge of towns.
- It will have to have **access** to roads.

Finding a suitable site

1 A factory

Look for a site that has:

- **Space** for deliveries, storage and workers' cars.
- **Access** to main roads for bringing goods to and from the factory.

- A site at the edge of town because land is cheaper and a factory may need a lot of space.

2 A new school

Look for a site that has:

- **Space** for a playground or playing fields.
- **Access** to roadways for staff cars and deliveries.
- A housing estate within walking distance.

How to present your answer

When giving reasons why you are picking a particular location for the activity follow this example. The colours will help you remember the three steps.

- Name the location.
- State the reason.
- Explain the reason.

Junior Certificate Geography Exam
Question
Find a place on the photograph of Westport where you would like to live. Give a reason for your answer.
Answer

1. I would like to live in the small housing estate in the left background. This is a small estate of bungalows in a cul-de-sac. This would be a quiet and safe place to play.

Aerial view of Westport

Traffic in photographs

1. Identify a place in the photograph of Westport where it is likely that **traffic congestion** would happen.
2. Give examples that efforts have been made to manage traffic flow in the town.

Look at the picture and find the following:

Congestion points
- Where a number of roads meet.
- Where a road narrows and a number of lanes feed into this narrow spot.
- Where traffic feeds onto a bridge.

Signs of traffic management
- Car parking spaces.
- Yellow boxes at road junctions.
- Arrows showing one way streets.
- Yellow lines along kerbs.

Key points

- **Space** and **access** are key points to consider when choosing a site for an activity.
- Aerial photographs can show where traffic congestion is likely to happen.
- Aerial photographs show signs of traffic management in towns.

Coastlines and Countryside on Photographs

Key words

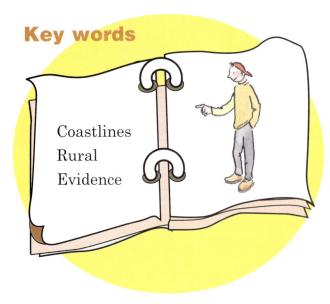

Coastlines
Rural
Evidence

Photographs of rural areas

So far we have looked at photographs of towns. Let's look at two other types of photographs:

- Photographs of **coastlines**.
- Photographs of **rural** areas (the countryside).

Many photographs combine landscapes. You may see part of a town and part of the countryside in a photograph.

Human activities along coastlines

When talking about human activities on coastlines you are looking for **evidence** of how people use the coast. Human activities include:

- Fishing
- Trading
- Tourism and recreation
- Defending the coast against erosion.

Example 1

Sample question

Look at photograph A. Name and explain one piece of evidence that this coast is popular for leisure activities.

Sample answer

Boating is a popular leisure activity in this place. There are small boats moored at the pier in the right centre of the photograph. They appear to be yachts as I can see sailing masts.

Aerial view of Kinsale

Example 2

Sample question

Look at photograph ⚠B. Name and explain one piece of evidence that this is a trading port.

Sample answer

There are large ships docked along the quay in the centre of the photograph. Goods would be loaded and unloaded here and stored in the large warehouses along the quay.

Photographs of rural areas

Rural refers to the country. These are the places outside the town. The activities in rural areas are mainly farming and forestry.

Farming

Look at photograph ⚠C. It shows what kind of farming happens in this area.

Trees are grown on higher land.

No crops or trees can survive on high and rocky land.

Crops: different coloured fields and lines of crops suggests cereals or vegetables.

Pastureland: this land is used to grow grass.

This photograph was probably taken in summer. The trees are green and the fields are golden.

Key points

- Aerial photographs show evidence of the work of the sea.
- Human activities such as trading and leisure activities can be seen on a photograph.
- Different farming activities and the time of the year can be read from a photograph.

Linking Maps and Photographs

Key words

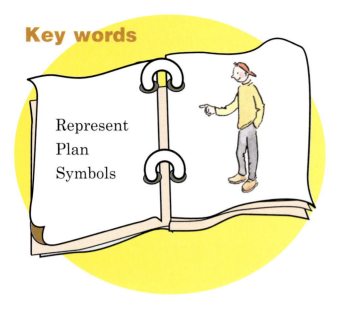

Represent
Plan
Symbols

Look at map △A and photograph △B of Tramore.

Maps and photographs may look very different but the general purpose of each is the same. The purpose of each is to **represent** a place.

● A photograph is a picture of the actual place.

● A map is a **plan** and uses **symbols** to show the features of a place.

Let's look at the main differences between a map and a photograph in more detail.

Maps

● The exact location of places is given using grid references.

● The scale of the map is known so distances can be worked out.

● The height of the land is known because it is represented in colour or by using contours or spot heights.

● The roads in an area are named although not all the streets in the town are shown on the map.

● Towns are named on the map.

Photographs

● The location is not exact. You can only tell where places are in relation to each other (beside, near, across etc.).

● It is difficult to work out distances between places.

● It is not possible to know the height of the land or the slope of it.

● You can see the street pattern in a town and that some roads are wider than others.

● You may be able to see differences in the type of houses between one part of the town and another.

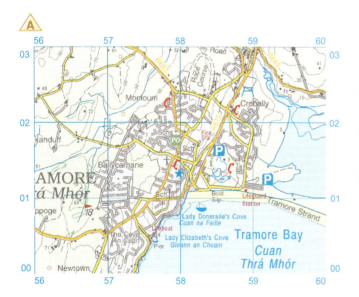

Using maps and photographs together

When you use the map and the photograph together, you can arrive at a good understanding of a place and what is going on there.

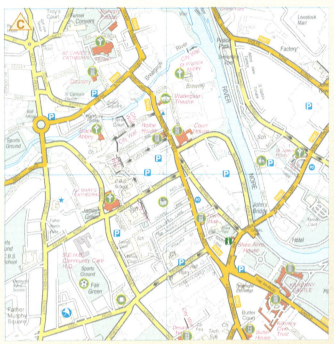

Map of Kilkenny

Aerial view of Kilkenny

Tourism on maps and photographs

Sample question

Using map C and photograph D describe the main tourist attractions of this area.

Sample answer

Tourists will be interested in historical attractions and leisure activities.

Historical attractions: They come to see the history of a place. There is a castle to see and town walls to view. The map has evidence in the bottom right-hand grid square. The photograph shows a castle in the north-west of it.

Leisure attractions: The photograph shows that there are gardens to relax in at the left background of the photograph.

Key points

- Maps and photographs when used together give us a good idea of the activities going on in an area.
- Maps and photographs differ in the way they **represent** or show a place.

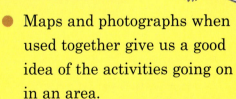

REVISION EXERCISES

Write the answers in your copybook.

1   Look at the photograph of Tramore on p. 248. Where would you find a car park?
    - Right foreground
    - Centre foreground
    - Centre
    - Left background

2   An aerial photograph that shows some sky is:
    - A low oblique photograph
    - A high oblique photograph
    - A vertical photograph

3   Draw a sketch of the photograph of Trim on p. 268 and mark in and name the following:
    - Shops
    - A castle
    - A row of houses
    - Two connecting roads
    - A river
    - Two bridges

4   Look again at the photograph of Trim on p. 268. Buildings in a town are used for different purposes. Identify and explain **three** services offered in this town.

5   Look at the photograph of Buttevant on p. 269. Imagine it has been suggested to build a new shopping centre at the site of the green field in the left background of the photograph.
    (a) Explain **one** reason why some people might be in favour of this site.
    (b) Explain **one** reason why some people might be against this site.

6   Study the photograph of Buttevant on p. 269. At what time of year was this picture taken? Using evidence from the photograph, give **one** reason for your answer.

7 *The main street in Buttevant runs across the photograph from left foreground to right background.*

  (a) With reference to the photograph on p. 269, give **two** pieces of evidence to support the statement that this is Buttevant's main street.

  (b) Briefly explain **one** advantage and **one** disadvantage in living along the main street of a town like Buttevant.

8 Identify a location in the town of Sligo where you think traffic congestion might occur. Explain **two** reasons why you selected that location.

Sligo

9 Identify **two** pieces of evidence on the photograph of Sligo that show how traffic is managed or controlled.

10 Study the photograph of Sligo town. Choose and clearly identify a place on the photograph where you would like to live. Give **three** reasons for your answer using details from the photograph.

11 Imagine you are a member of Sligo Corporation. You are at a meeting that is discussing spending money to improve the town. Describe **two** things you would suggest. Use as much detail from the photograph as possible to back up your views.

24 Sample Answers

In this section we will look at how to answer questions in the Junior Certificate examination. We will focus on the type of questions given in **Section Two** of the examination paper.

- Begin by reading all five questions. Select your **three** best questions. Before you write in your answer book make quick notes on the paper to test whether you can answer all parts of the question. You may need to choose a different question.

- Note the number of marks given for each answer. It is given as a number in brackets beside the question. This tells you how much time you should give to answering that part. It also tells you how much information you need to give. **Two marks** tell you that you are expected to give just one point. **Don't waste time giving more.**

- Where possible use the question to help you with your answer:
  Example: Describe one effect of overfishing in the seas around Ireland.
  Answer: One effect of overfishing in the seas around Ireland is...

Answering skills

It is useful to have a clear way of answering the more detailed questions i.e. the questions offering more marks. One method to apply is **PEN**:

P: **P**oint   E: **E**xplanation   N: **N**amed example

Note: *It may not be necessary/possible to give a named example in all cases.*

Example: Why do tourists go to Spain?

Answer: Tourists go to Spain because the summers are always warm and sunny **(Point)**. Visitors to Spain know that there are over 320 days of sunshine there each year. Many come from colder areas like Ireland. They want guaranteed sunshine that will allow them to enjoy the sea and the many services that Spain offers **(Explanation)**. Popular resorts include Marbella **(Named example)**.

Sample answers
2004 Ordinary Level Question 1, Section 2

2.   **USE OF RESOURCES**

A. WATER

Look at the diagram above. It shows seven sources of river pollution.

Choose **THREE** of these. For each of the ones you have chosen:
*   Say what is causing the pollution.
*   Describe how it is a problem.   (9)

Sample Answer

Three sources of pollution shown in the diagram are:
*   Farm spills
*   Power stations
*   Waste barrels

The pipe may be draining farm spills **(Point)**. This can kill fish and plant life in the river **(Explanation)**. Examples of farm spills are fertilisers and pesticides **(Named example)**.

The hot water from the power station means there is less oxygen in the water **(Point)**. Fish cannot then survive **(Explanation)**.

Spills from waste barrels may pollute the water **(Point)**. Oil pollution is dangerous for people swimming in the river **(Explanation)**.

B. FARMING

The photograph shows a farmer spraying barley.

Describe the farm as a system, referring to **TWO** inputs, **TWO** processes (apart from the spraying shown) and **ONE** output. (10)

Sample Answer

A farm is a system. **Two** inputs are fertile soil and machines. **Two** processes are ploughing the land and sowing the seeds. **One** output is cereals **(Point)**. During autumn the farmer ploughs the field getting the soil ready for spring. In spring the farmer will sow the seeds. Then in late summer when it is dry he/she will harvest the crop **(Explanation)**. One such crop is barley **(Named example)**.

C. FISHING

   (i)   Overfishing has caused a shortage of fish in many sea areas.
        Explain **TWO** ways more modern fishing boats and equipment have led to overfishing. (6)

  (ii) Describe **ONE** effect of overfishing in the seas around Ireland. (3)

 (iii) Explain what is meant by a **renewable resource**. (2)

Sample Answer

   (i)   Modern boats have led to overfishing as they use a lot of technology **(Point)**. One way is using sonar equipment, which hears the fish sounds. This helps to locate the fish. They cannot escape. A second way is by having large refrigerators, which allow the boats to stay out at sea for longer periods. This means more fish can be caught **(Explanation)**. The trawlers at Killybegs, Co. Donegal are modern **(Named example)**.

(ii) One effect of overfishing in the seas around Ireland is that herring stocks decreased **(Point)**. Serious measures such as quotas have been set **(Explanation)**.

(iii) A renewable resource is something that is useful to humans and can be used again and again.

2003 Ordinary Level Question 2, Section 2

2. ## PHYSICAL GEOGRAPHY

A.

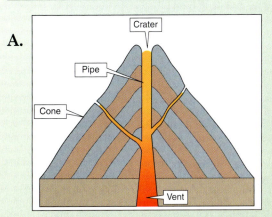

Look at the diagram. Describe how a **VOLCANO** is formed. In your answer use the names from the diagram. (10)

Sample Answer

A volcano is formed when magma from beneath the crust rises through a vent **(Point)**. It flows out onto the earth's surface through a vertical passageway called a pipe. In time it builds up into a cone-shaped mountain. At the top of this mountain is a wide, bowl-shaped hollow called a crater **(Explanation)**. Mount St Helen in Washington state, USA formed in this way **(Named example)**.

B. (i) Describe **THREE** types of damage caused when a volcano erupts. (6)

Sample Answer

When a volcano erupts many lives can be lost **(Point)**. The hot gases and ash can mix with ice on the peak of the volcano causing a mudflow **(Explanation)**. The 1984 eruption of Nevado del Ruiz in Colombia, South America, killed over 21,000 people **(Named example)**.

Volcanoes can destroy property **(Point)**. Millions of tonnes of lava can run down a mountainside covering houses and other buildings. People are left homeless **(Explanation)**. The 1983 eruption of Mount Etna in Sicily destroyed many houses **(Named example)**.

Volcanoes can destroy crops **(Point)**. When an eruption happens ash and other materials can cover or burn the crops. This can seriously harm people's livelihoods **(Explanation)**. The 1983 eruption of Mount Etna destroyed many orange groves and vineyards **(Named example)**.

(ii)   Explain **TWO** ways volcanoes can be useful to people. (4)

Sample Answer

Volcanoes produce very fertile soils **(Point)**. This means farming produces high crop yields **(Explanation)**. An example of this is coffee grown in Brazil **(Named example)**.

Volcanoes attract a large number of tourists **(Point)**. This brings money and jobs to the local area **(Explanation)**. One volcano that attracts many tourists is Mount Vesuvius in Italy **(Named example)**.

C. **ROCKS** can be used in many different ways by people.

(i)   Name **THREE** rocks and say how each can be used.   (6)

(ii)   Explain why people might object if a quarry to extract rock was opened near their home.   (4)

Sample Answer

(i)   Three rocks are granite, limestone and marble. *(NB: You are asked to name the rocks only).* Granite can be used for buildings, as it is hardwearing. Limestone can be crushed to make cement. This is useful for the building industry. Marble is used for kitchen tops and fireplaces in homes.

(ii)   People might object because of the increase in traffic and noise pollution **(Point)**. The heavy trucks that transport the products and the loud blasting equipment could upset locals **(Explanation)**. One such quarry is at Belgard in Tallaght, Co. Dublin **(Named example)**.
You may notice that in (i) above the answers were very short. This was based on the marks given.

1. **URBAN GEOGRAPHY**

A. Examine the table of Ireland's fastest growing urban centres (excluding Dublin).

The table shows the percentage change in population between 1981 and 1996.

Centre	Change %
Bray	22.2
Lucan	-
Swords	100.3
Navan	15.0
Newbridge	24.7
Leixlip	44.5
Malahide	47.8
Arklow	-1.0
Naas	68.7
Portmarnock	11.4
Greystones	34.3
Balbriggan	26.3
Skerries	26.7
Athy	-4.7
Wicklow	36.5
Celbridge	168.1
Kildare	6.5
Rush	40.5
Kells	-3.3
Trim	24.9
Maynooth	151.7
Ashbourne	115.0

(i) Put in rank order the **four** fastest growing urban centres. (4)

(ii) With reference to **one** Irish urban centre you have studied explain the factors which led to its development.

In your answer refer to:
* Economic factors
* Social factors
* Administrative factors (12)

Sample Answer

(i) The four fastest growing urban centres in rank order are Celbridge, Maynooth, Ashbourne and Swords.

(ii) The urban centre I would like to discuss is Dublin city. Dublin is a port at the mouth of the Liffey **(Point)**. Most of the country's trade passes through this port. As the port grew the economy grew. Money was invested in building an excellent transport network linking the port to all parts of the country. Goods and people could then be moved easily along roads and railways **(Explanation)**. Industries such as Guinness provided jobs **(Named example)**.

Dublin's social life offered a variety of activities, which has attracted people to the city **(Point)**. The many theatres, night-clubs, concerts and restaurants brought a good social mix. This social mixing led to more marriages and births, which in turn led to more people settling in the city **(Explanation)**. Dublin's social facilities include the Abbey Theatre and Croke Park **(Named example)**. The centre of Ireland's administration is located in Dublin **(Point)**. People move close to the centre of decision-making where jobs are more available. Many embassies have also located in the city. The number of services available grew **(Explanation)**. The Dáil in Dublin is the centre of Ireland's administration **(Named example)**.

B. Examine the diagram showing urban development and answer the questions that follow:

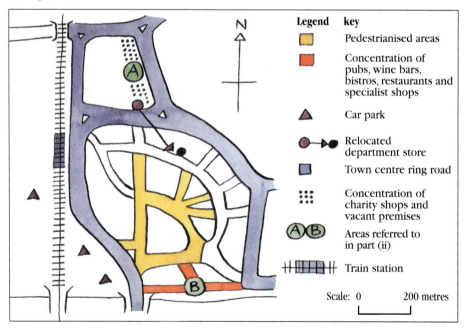

(i) What evidence is there that the local council has seen traffic management as an important element in the planning of this town centre? (4)

(ii) Suggest **two** reasons why the area marked **B** appears to be developing while the area marked **A** appears to be in decline. (4)

(iii) In **any** town which you have studied, explain **one** recent development which helped to solve the problems of urban sprawl **or** urban renewal/ redevelopment. (6)

(30 marks)

Sample Answer

(i) There is a ring road that circles the town **(Point 1)**. This allows cars to by-pass the town centre. Traffic can be managed more carefully by the local council **(Explanation 1)**. The local council have allowed three major new car parks to be built on the edge of the town **(Point 2)**. Fewer cars enter the town centre. There are fewer traffic jams there **(Explanation 2)**.

(ii) Area B has a concentration of pubs, wine bars, bistros, restaurants and specialist shops. This is a sign of growth. Area A has a concentration of charity shops and vacant premises. This is a sign of decline **(Point 1)**. People like to spend their free time in an attractive, lively area with lots of social activities. Once it is seen as an attractive lively location other services will set up close by. There will be more growth. A department store has relocated from Area A to Area B. Areas with charity shops and vacant sites do not bring growth **(Explanation 1)**.

Area B is closer to the pedestrianised area **(Point 2)**. Pedestrianised areas are areas where there is a high volume of people passing by. They are pedestrianised when there is seen to be a lot of business in the area. Areas with vacant sites suggest that the demand for the site has declined **(Explanation 2)**.

(iii) The town that I have studied is Dublin. One recent development that has helped to solve the problems of urban renewal/redevelopment is the development of housing in certain areas **(Point)**. When an area has been redeveloped offices and shops are often its only functions. This means that at night the area is empty and vandalism can occur. By encouraging people to live and work here there is less crime in the area **(Explanation)**. One such development took place in the Docklands area of Dublin **(Named example)**.

Maps

Political Map of the World

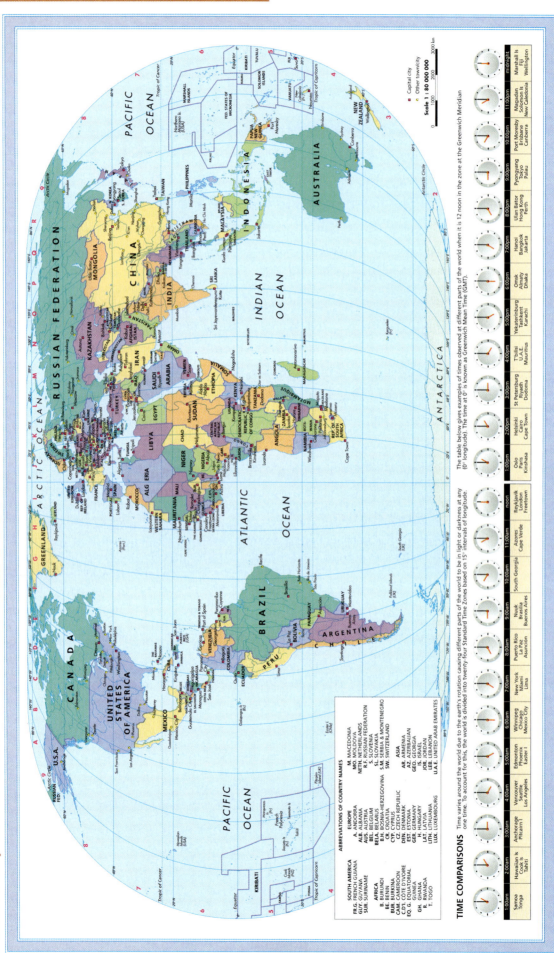

Glossary

Abrasion: A process of erosion that uses rocks to wear down other rocks.

Absentee landlords: People who own land or property but do not live or work on it.

Accessibility: How easy a place is to get to.

Administrative: The management or control of a company or country.

Aerial photograph: A photograph of an area taken from an aeroplane. Can be vertical or oblique.

Agricultural products: Goods that come from farms.

Bilateral aid: Help that is given by a government of a rich country to a government of a poor country.

Birth rate: The number of people born in one year for every thousand people in the population.

Blanket bog: Areas where thin layers of wet peat form a blanket over the landscape.

Bridging point: The narrowest point on a river where it is best to build a bridge.

Business parks: Sites on which a number of businesses set up close together.

Bustees: Shanty towns in India.

By-pass: A routeway that avoids a town by going around it.

Carrying capacity: The ability of an area to support its people with food and jobs.

Census: A head count of the people living in an area.

Central business district (CBD): The centre of a town or city where offices and shops are found.

Colonial power: The country or people who take over another country and control it.

Colony: A country or region that is taken over and exploited by another country.

Commodities: Goods that are bought and sold on the world market.

Community schemes: Plans or projects set up by people in their local area e.g. schools and hospitals.

Commuter: Someone who travels some distance to get to and from work.

Conflict of interest: This refers to a disagreement between people on how to use a resource.

Conservation: The ways of using resources wisely and protecting the environment.

Conurbation: When towns or cities keep spreading out until they meet or melt into each other using up the green space in between.

Co-operative: A business that is set up and owned by a group of people.

Credit loans: Loans, usually of money, that are given out to help people.

Death rate: The number of people dying in one year for every 1,000 people in the population.

Decentralisation: Movement of people and services from a central area like a large city to a local area.

Demography: The study of population numbers.

Derelict: An abandoned house or site.

Designated areas: Areas that are chosen for the development of houses, offices and factories and which are given tax breaks and grants.

Destination area: The area to which migrants move.

Detached housing: Houses that stand on their own and are not attached to another house.

Developed country: A country that has a lot of money, many services and a high standard of living.

Developing country: A country that is often quite poor, has few services and a low standard of living.

Development: How rich or poor a country is compared with others.

Development aid: Long-term help that is given by richer countries to poorer countries. This type of aid includes school and hospital projects.

Dispersed settlement: Areas in which the houses and farm buildings are scattered over a wide space.

Dry cattle: Cattle that are reared to produce beef.

Dry-point: A site on high ground near a river that can flood from time to time.

Dyke: A ditch or a bank around a low-lying area.

Economic activity: An activity from which people make a living. This can be a primary, secondary or tertiary activity.

Economist: A person who studies the wealth of a country and its people.

Economy: The wealth of a country.

Electronic connections: Links, such as the Internet, that require electricity to operate.

Emergency aid: Short-term help given by richer countries to poorer countries. This type of aid includes food, medicines and blankets.

Emigration: The act of leaving the home country to live in another country.

Energy: A force for heat or movement.

Environment: The natural or physical surroundings where people, animals and plants live.

Exploit: To develop or use a country's resources to the user's advantage.

Exports: Goods sold to other countries.

Extractive: Taken out of the ground.

Favela: The Brazilian name for a shanty settlement.

Finite: A resource that is limited and will run out when it is used.

Footloose: An industry that is not tied to a particular location.

Fossil fuels: Products such as coal, oil and gas that formed millions of years ago from the remains of animals and plants.

Function: The main purpose of a town. Functions include industries, education and health.

Geothermal power: Energy obtained from the hot rocks found under the earth's surface.

Global village: Modern communications such as the Internet or the telephone have made it easier to be in touch. The phrase suggests the world has become a smaller place.

Gross National Product (GNP): The wealth of a country – its total income divided by its total population.

Green belt: A protected area of countryside around a city. New building is controlled to try and stop the spread of the city.

Greenheart: An area of open space in the middle of a built up area.

Grid references: A group of four or six figures used to find a place on an Ordnance Survey map.

Hay: Grass that has been cut and dried.

Hinterland: The area around a town served by the services (shops, schools) in that town.

Horizon: The furthest point you can see where the earth meets the sky.

Horticulture: The farming of fruit, vegetables and flowers.

Hydroelectric power: Electricity obtained from using fast flowing water.

Illiteracy rates: The percentage of people who can neither read nor write.

Immigration: The act of coming into a country from another country.

Immunisation: The method of protecting or keeping people free from diseases.

Incentives: Ways of encouraging people to act, e.g. giving grants.

Industrial decline: This is when an industry is dying or is making less profit.

Industrial inertia: This is when an industry fails to move location even though it may be more profitable for it to do so.

Industrial location: The site where an industry is found.

Industrially emergent: Places that are attracting more industries.

Industry: Commercial production and sale of goods.

Infilling: Building of new houses in gardens and empty sites.

Information superhighway: This refers to the way in which information and knowledge is now available along route ways such as the world wide web (WWW).

Information technology: These are machines, like the computer, that provide information.

Infrastructure: The links of communication like transport routes and telephone lines found within an area.

Inner city: An area of factories and old houses next to the city centre.

Inputs: The things that are needed when setting up an activity e.g. raw materials and money.

Insecticide: A poison that kills insects.

Irrigation: The artificial watering of an area by sprinklers etc.

Key: A list of signs and symbols on a map or diagram with an explanation of what they mean.

Landlocked: A country that is surrounded by other countries and has no boundary with the sea.

Leaching: A process in soils where water brings minerals downward in the soil leaving the top layer infertile.

Life-expectancy: The average number of years a person can expect to live.

Linear settlement: Places where houses are spread out in a line along a main road, a river or a coastline.

Locational factors: The reasons why factories set up on a particular site.

Manufacturers: Companies who change a raw material into finished goods.

Manufacturing: A secondary industry where goods are made.

Manure: Animal dung that is used to fertilise soil.

Map: A drawing which shows part of the earth's surface on a reduced scale.

Market: A group of people who buy raw materials or goods.

Market town: A town where people could buy and sell goods. This was the original function of many towns.

Meadow: Open fields where animals graze on grass.

Medieval: Dating to the Middle Ages.

Migrants: People who move from one place to another to live or to work.

Migration: The movement of people from one place to another to live or to work.

Monitor: To keep a close look and check on something.

Multifunctional: Something that has many purposes or reasons for existing.

Multilateral aid: Help that is given by governments of rich countries to a central agency like the U.N. who decide in which poorer country aid is most needed.

Multinational companies: Businesses that have branches in many countries.

National park: An area of countryside preserved by law from development taking place.

Natural change: The way a population changes as a result of its birth rate and death rate. Migration rates are not included.

Natural decrease: The death rate is greater than the birth rate so the population falls.

Natural increase: The birth rate is greater than the death rate so the population grows.

Natural resources: Raw materials that are obtained from the environment e.g. soil, water, oil, gas.

Network: A number of lines that connect with each other.

Nodal point: A point or area where many lines of communications meet, e.g. at a crossroads.

Non-governmental aid: Help that is given by voluntary organisations to people in poorer countries e.g. Self Help.

Non-renewable: Resources that can only be used once.

Nucleated settlement: Places where houses are grouped closely together.

Optimistic view: A view that sees the brighter side of things, for example, that population will not continue to rise.

Ordnance Survey: The government organisation responsible for producing maps.

Outputs: The finished goods from an activity e.g. farming or manufacturing.

Overpopulated: When there are more people living in an area than the area can support.

Pasture: Areas of grass on which animals feed.

Pattern: How things are spread out over an area of land.

Pessimistic view: A view that sees the darker side of things, e.g. that population will continue to rise causing famines and other disasters.

Planned towns: Towns where the use of land was carefully planned.

Plantations: Large farms that usually grow one crop for export.

Polder: An area of land reclaimed from the sea.

Pollution: Noise, dust, chemicals in water and other harmful substances that are produced by people and machines and can spoil an area.

Population: The people who live in an area.

Population density: The average number of people living in a square kilometre of space.

Population distribution: How people are spread out over an area.

Population explosion: A sudden quick rise in the number of people.

Population growth: An increase in the number of people in an area.

Population pyramid: A bar graph which displays the age and sex make up of a population.

Preservatives: Anything that is added to a product to keep it from rotting.

Primacy: Refers to those cities such as Dublin or Paris where the population is over twice the size of the next largest city in the country.

Primary activity: An activity that involves natural resources e.g. farming, fishing, forestry and mining.

Primary goods: Raw materials such as foodstuffs, minerals and timber. They are usually of low value and give less profit to the producer than manufactured goods.

Processes: The ways of changing raw materials into something else.

Pull factors: Things that attract people to live in an area.

Push factors: Things that make people want to leave an area.

Quality bus corridor: This is a lane of a main road that is for bus and taxi use only.

Quality of life: How happy people are with their environment and what it has to offer them.

Quotas: A fixed share of something.

Raised bog: This is a large low-lying area of deep peat.

Rationalised: The cutbacks that are made when a factory is not making enough profit.

Raw materials: Natural resources that are used to make things.

Reclaimed land: Land that has been recovered from the sea and can be put to some use.

Redevelop: To develop a site or area in a new way. It often means demolishing older buildings.

Remittances: This is money that is sent back from emigrants to their family.

Renewable: A resource that can be used over and over again.

Reservoir: An artificial or man-made lake used to store water.

Resources: A useful supply that can be drawn on when needed. Resources can be natural e.g. gold and silver or man made e.g. educated, skilled people.

Retailers: People who sell goods in shops.

Retail price: The price that shops charge for goods.

Rural: An area of land which is mainly countryside.

Rural-to-urban migration: The movement of people from the countryside to the towns and cities.

Satellite towns: Towns that are found close to a larger city and who depend heavily on that large city for its jobs, social life etc.

Scale: The link between the distance on a map and its real distance on the ground.

Scale line: A line on a map which shows how far real distances are.

Secondary activities: Activities that turn natural resources into goods that we can use. Secondary activity is another term for the manufacturing industry.

Semi-detached housing: Houses that are built in groups of two, which are attached to each other on one side.

Settlement: A place where people live.

Shanty town: An area of poor quality housing which often lacks basic services such as electricity, a water supply and sewage disposal.

Silage: Grass that is cut and covered in plastic and is later fed to animals.

Site: The actual ground where a settlement or industry is built.

Slurry: Animal waste that is collected in drains and used as a fertiliser.

Solar power: Electricity that is obtained from trapping the sun's heat.

Source area: The area from which people migrate.

Spot height: A point on a map with a number giving its height above sea level in metres.

Standard of living: How well off a country or person is.

Subsidies: Money that is given as a help or support to people.

Sustainable development: A way of making progress that does not waste resources. It looks after the needs of today without damaging resources for the future.

Tariffs: Taxes that are put on goods when they are imported or exported.

Technology: Machines or tools that require skill and knowledge to use.

Terraced housing: Houses that are built in groups and attached to each other.

Tertiary activities: An industry that provides a service for people such as teaching and nursing.

Tidal power: Electricity that is obtained from trapping the energy of waves.

Tied aid: Help that is given by rich countries to poor countries, but with conditions attached.

Trade: The movement of goods and services between countries.

Trend: When something tends to go in a particular direction forming a pattern.

Urban: A town or city.

Urbanisation: The growing proportion of people living in towns and cities.

Vaccines: Medicines that protect against diseases.

Wind power: Electricity that is obtained from trapping the power of the wind.

Zones: Areas with similar functions or features.

Aerial photographs

Aerial view of Trim

Aerial view of Buttevant